The Latin American Studies Book Series

The Latin American Studies Book Series promotes quality scientific research focusing on Latin American countries. The series accepts disciplinary and interdisciplinary titles related to geographical, environmental, cultural, economic, political, urban and health-related research dedicated to Latin America. The series publishes comprehensive monographs, edited volumes and textbooks refereed by a region or country expert specialized in Latin American studies.

The series aims to raise the profile of Latin American studies, showcasing important works developed focusing on the region. It is aimed at researchers, students, and everyone interested in Latin American topics.

Submit a proposal: Proposals for the series will be considered by the Series Advisory Board. A book proposal form can be obtained from the Publisher, Andrea Brody-Barre (andrea.brody-barre@springer.com).

Leonidas Osvaldo Girardin

Socioeconomic and Geopolitical Aspects of Global Climate Change

An Intersectorial Vision from the South of the South

Springer

Leonidas Osvaldo Girardin
CONICET/Fundacion Bariloche
Environmental Department
Buenos Aires, Argentina

ISSN 2366-3421 ISSN 2366-343X (electronic)
The Latin American Studies Book Series
ISBN 978-3-031-53245-0 ISBN 978-3-031-53246-7 (eBook)
https://doi.org/10.1007/978-3-031-53246-7

This Springer imprint is published by the registered company Springer Nature Switzerland AG
The registered company address is: Gewerbestrasse 11, 6330 Cham, Switzerland

Paper in this product is recyclable.

Contents

1 Introduction. Socioeconomic and Geopolitical Aspects of Global Climate Change. An Intersectoral Vision From the South of the South. Situation of the International Negotiation from the Paris Agreement (December 2015) and Its Latter Evolution. "We Were Few, and the Pandemic Arrived" 1
References 11

2 Agriculture and Climate Change: Opportunity or Threat? 15
References 29

3 Climate Change and Semi-arid Regions in Latin America Threats and Challenges 31
3.1 Introduction 31
3.2 The Climate Change Context 35
3.3 Climate Change and Agriculture 36
3.4 Climate Change and Agriculture in Latin America 41
3.5 The Gran Chaco Region (Argentina-Bolivia-Paraguay) 47
3.6 The Semi-arid Region in Brazilian Northeast (the Sertão) 63
3.7 The Venezuelan Coastal Zone 68
3.8 The Semi-arid Belt of Central America 73
3.9 Conclusions and Recommendations 91
References 96

4 Energy and Climate Change: Challenges for Low-Carbon Development 101
4.1 The Context of International Negotiation and Heterogeneity 102
4.2 Equity and Efficiency. Future and Current Generations 107
4.3 Energy and Climate Change 111
4.4 The "Inertia in Energy Consumption" and Future Perspectives According to the "Paris Agreement" and the Fulfillment of Its Goals 120

4.5 Challenges Related to Sustainability and the Role of the Energy Sector 127
References 131

5 Socioeconomic and Political Aspects of Climate Change. The Role of the Clean Development Mechanism and Other Market-Based Mechanisms in Contributing to the Ultimate Objective of the UNFCCC and Sustainable Development. A Latin American Point of View of the Situation After the Paris Agreement 135
5.1 Introduction 136
5.2 Latin America: From Euphoria to Disenchantment 137
5.3 Climate Change and Heterogeneity 138
5.4 Resource Allocation and Climate Change (I): Who Pays and Departing from Which Argument? 139
5.5 Resource Allocation and Climate Change (II): Synergies and Conflicts Between Adaptation and Mitigation 141
5.6 The Role of the Market and the CDM in Contributing to Sustainable Development: "From Saying to Doing ..." 143
5.7 CDM: What Could Be Expected After Doha? 147
5.8 The Paris Agreement, the Poor Results of COP25, Trump and Beyond 148
References 151

6 Regional Study on the Economics of Climate Change in South America. Argentine Chapter (ERECCS-Argentina). ECLAC 155
References 160

7 Forest Fires in Australia: Are We Inevitably "In the Oven" Also in Argentina? 161
References 169

About the Author

Leonidas Osvaldo Girardin is Economist and has master's degree in Environmental and Territorial Policies (University of Buenos Aires), master's degree in Economics with orientation in Economic Development, International Trade and Employment (Di Tella University, Argentina), and he has post-degree studies in Environmental and Energy Economics (University of Comahue, Argentina). He is Researcher at the Argentine National Council on Science and Technology (CONICET) and Senior Researcher at the Bariloche Foundation. His main fields of activity are environmental economics, climate change economics, and the relationship between development, economy, energy, and environment. He acted as Consultant at ECLAC, World Bank, IADB, UNDP, UNEP, among other international institutions. He is Regular Professor at many universities not only in Argentina but also in other Latin American countries. He was Member of the IPCC Bureau of the Task Force on Inventories, from 2008 to 2015, and the UN Expert Group on Technology Transfer from 2002 to 2005. From 1996 up to now, he integrates the Roster of Experts of UNFCCC for reviewing National GHG Inventories, National Communications, and Biennial Reports of Annex I Parties.

Chapter 1
Introduction. Socioeconomic and Geopolitical Aspects of Global Climate Change. An Intersectoral Vision From the South of the South. Situation of the International Negotiation from the Paris Agreement (December 2015) and Its Latter Evolution. "We Were Few, and the Pandemic Arrived"

Abstract This chapter explains the main socioeconomic and geopolitical aspects related not only to the international negotiation process on climate change, but also to some fundamental concepts for understanding the difficulties of reaching an agreement in this negotiation, which are presented in the introduction. Heterogeneity is a key factor in explaining the existence of these difficulties, insofar as there are marked differences in the geographical distribution of expected impacts, the degrees of existing vulnerability, the capacity to face the challenges posed by climate change, or the incidence of economic impacts linked to the measures taken to mitigate or adapt to climate change, but also between the current and historical responsibilities for having reached the current situation.

Keywords Climate change negotiations · Climate change socioeconomic issues · Climate change geopolitical issues · Climate change economics · Historical responsibilities · Common but differentiated responsibilities

The international negotiation process related to climate change issues is extremely dynamic, with procedures, modalities, and topics of discussion changing rapidly over time. Sometimes these changes occur from meeting to meeting. However, the basic problem remains unchanged: how to reduce and/or limit anthropogenic greenhouse gas (GHG) emissions to a level that does not interfere with natural and human climate systems (or, in other words, is not dangerous for human life and the rest of the species that inhabit the Earth), and, at the same time, how to share the costs of mitigating these emissions and adapting to the impacts that cannot be avoided?

It comes from the phrase "We were few and grandmother gave birth" is an expression used in those situations where something bad happens and things get worse unexpectedly.

L. O. Girardin, *Socioeconomic and Geopolitical Aspects of Global Climate Change*,
The Latin American Studies Book Series, https://doi.org/10.1007/978-3-031-53246-7_1

Since the signing of "Paris Agreement"[1] a new question arises, and it is whether the Nationally Determined Contributions (NDCs) announced by the countries that signed the Agreement will be a real contribution to the fulfillment of the objective of not increasing the average temperature of the planet above 2 °C (and even more to the additional effort proposed of not exceeding 1.5 °C) toward the end of the twenty-first century[2] or if, in short, this does not result in a new postponement of the decision to face the problem.[3,4] A problem whose serious treatment, in fact, has been postponed since the signing and ratification of the United Nations Framework Convention on Climate Change (UNFCCC) almost 30 years ago.

As if this challenge of meeting the ultimate objective of the UNFCCC,[5] which seems increasingly complex, is not enough, since the beginning of 2020 a new uncertainty is added: what consequences will be brought in the short, medium, and long term by the COVID-19 pandemic that determined a strong change in social behavior, economic activities, and trends in GHG emissions across the planet.[6]

The way in which these questions are resolved is not trivial, since it is increasingly clear that, beyond the efforts made to mitigate GHG emissions (and consequently their atmospheric concentrations), it will also be inevitable to suffer some kind of impact of a certain magnitude, mainly (and paradoxically) in the countries and regions that were not necessarily the ones that historically have contributed the most to the problem.[7] In this sense, the first results of the Studies on the Economics of Climate Change in Latin America and the Caribbean carried out by ECLAC show that for the countries of the region (even if we take into account those of greater importance, such as Argentina, for example), it is necessary to take into account the fact that the region has been affected by climate change for many years, (Brazil and/ or Mexico); the costs of potential impacts of climate change and/or variability, which are expected to occur with a higher degree of probability in the future (from the most likely climate scenarios), are extremely high and lead to the need to allocate huge amounts of resources to adapt to these new situations. On the other hand, the costs of potential mitigation measures can be comparatively very important for the size of

---

[1] Which is the document that emerges as the main result of the COP-21, held between November 30 and December 11, 2015, in that city.

[2] 2 and 1.5 °C above the records of the time before the Industrial Revolution.

[3] See, among others, Girardin (2000, 2008, 2013) and Girardin et al. (2014, 2017a, b, c). For a follow-up of the international negotiation process, the best source is the website of the United Nations Framework Convention on Climate Change (UNFCCC) Secretariat itself: www.unfccc.int.

[4] A guideline for this difficulty is given by the results of the so-called "Emission Gap Reports", carried out by UNEP, which determine the "Emission Gap" that needs to be reduced through additional mitigation efforts to the NDCs presented by the different countries to ensure that the trajectory leading to the fulfillment of the 1.5 and 2 °C target is followed. See UNEP (2019).

[5] The ultimate objective of the UNFCCC is precisely to achieve stabilization of GHG concentrations in the atmosphere at a level that would prevent dangerous anthropogenic interference with the climate system. See UNFCCC (1992, 1997).

[6] See CEPAL (2020a, b, c, d), CEPAL/OIT (2020), and Nature Climate Change (2020).

[7] IPCC (2013a, b, 2014a, b), Stern (2006a, b), CEPAL (2010), and Girardin et al. (2014). It is also interesting to see PNUMA (2019).

their economies even if they do not necessarily represent such a significant reduction in global GHG emissions.[8]

Fundamentally these situations are what explain that an analysis that tries to unravel the socioeconomic consequences of the expected impacts of climate change, of the measures to mitigate or adapt to it, and even of the actions to adapt to the mitigation and/or adaptation measures taken by other countries[9] will have many points of contact between one period and another (before and after the signing of the Paris Agreement), although not necessarily the same issues are addressed from the same starting point, as this is constantly changing.[10] However, just as there are issues that are presented as crucial at one point in the negotiation and quickly pass into the background and others that from one moment to the next become critical, when they were not previously on the agenda of the discussions, there are also issues that remain as fundamental points that remain in the discussion over time.[11] Nevertheless, since the beginning of this year, the potential effects of the COVID-19 pandemic constitute (for various reasons) an unavoidable point to take into account in terms of the future of the international negotiation and the response of the different "Parties" of the international agreements on climate change. In any case, although some work has begun to circulate, it is still too early to have a full idea of the real implications.

An important issue to take into consideration when addressing these issues is that any policy or concrete measure adopted, with the aim of achieving a reduction in GHG emissions, will mean some kind of impact on the various activities, sectors, regions, systems, and/or populations involved and, therefore, some kind of sacrifice on the economy of the societies that implement them. That is why one of the most conflictive points on the agenda of international negotiations on climate change is the distribution of the costs of mitigating its effects among the various countries.[12]

The problems that each society faces are different and so are the degrees of vulnerability to which they are subjected. Thus, the interests of the various actors involved may conflict, depending on the modality implemented to try to address the

---

[8] For more detail on the Regional Studies on the Economics of Climate Change in Latin America and the Caribbean (ERECC), see www.eclac.org. See also CEPAL (2010) and Girardin et al. (2014).

[9] In this sense, the issues related to measures that can be applied, in the field of international trade, using as an argument the need to comply with GHG emission mitigation policies or environmental protection strategies, are of particular importance. Among these measures are border taxes on the carbon content of products based on the calculation of a "carbon footprint", under standardized methods. These measures can seriously affect trade and, mainly, the exports of certain products from countries (mainly in developing countries, but also developed countries) which (given the high component of emissions linked to transport over the total "life cycle" emissions of many of these products) could be particularly harmful to those countries located further away from the main international markets. See Bouzas (2011) and also www.ambienteycomercio.org.

[10] An example of this is the signing of the Paris Agreement by 195 countries (or "Parties" to the United Nations Framework Convention on Climate Change).

[11] This can be clearly seen just by reviewing the documents produced at different moments of the negotiation. See: www.unfccc.int and www.ipcc.ch.

[12] Girardin (2000, 2008), Bhaskar (1995), Argawall and Narain (1991), Lipietz (1995), among others.

problem. This situation leads to the adoption of different approaches to address the issue.[13]

From the economic point of view, the solution that is finally adopted will not be neutral in terms of the effects it has on the distribution of income among the various countries, nor among the various social groups within the countries themselves. The different methodological approaches, regarding the costs to be faced for the reduction of GHG emissions, lead to different results, depending on the models used in the formulation and simulation of future scenarios and the assumptions on which they are based. The same applies to the methods used for the evaluation of the different policy alternatives to be applied.[14]

Because of the close relationship between the assumptions that underpin the logical structure of the models used and the results that those models arrive at, there is a wide margin of uncertainty about the true costs involved in mitigation measures for each of the participants. However, there is some consensus that the cost per ton of the first reductions is lower, but it will increase significantly as the more accessible options are exhausted. Thus, the crucial issue of conflict and negotiation will then be the strategy chosen by each country and how the costs will be shared among nations.[15]

The United Nations Framework Convention on Climate Change (UNFCCC) refers to "common but differentiated responsibilities", thus recognizing the shared responsibilities to reduce CO_2 emissions, but also the right of relatively less developed countries to increase their energy consumption in the development process. According to this principle of "common but differentiated responsibility", all countries should take measures to prevent damage to the atmosphere, but the initiative and the primary effort should come from the "industrialized countries, while urging them to take the lead in" combating climate change and its adverse effects.[16]

However, in the Kyoto Protocol, "market" criteria were incorporated to achieve the objectives of reducing at least 5.2% of GHGs emitted by Annex I countries, in the period 2008–2012, compared to 1990 emission levels. This introduction into the Protocol of the so-called "Mechanisms for Cooperation in the Implementation of the Protocol",[17] designed to enable the countries that have assumed obligations to comply with them at lower costs means that priority will be given, in practice, to carrying out the measures in those places where the costs of mitigation are lower,[18] although this is supported by the possibility of receiving investments in sectors that the host countries consider contribute to their sustainable development.[19] Not only

---

13 Idem.

14 Criqui and Kouvaritakis (1997) and Girardin (2000).

15 Criqui and Kouvaritakis (1997), Girardin (2000, 2008).

16 www.unfccc.int.

17 This was the name these mechanisms originally took, although they were later known simply as "Kyoto Mechanisms".

18 These mechanisms are Joint Implementation (JI), the Clean Development Mechanism (CDM), and Emissions Trading (ET).

19 In addition to the mechanisms that arise from the KP, there are also other market instruments in which the reductions and/or limitations of GHG emissions can be valued. There are several

was this "prioritization" ratified in the Paris Agreement (which returns to the idea of implementing a carbon market at the global level), but the principle of "common but differentiated responsibilities" also seems to be diluted, as all "Parties" to the Agreement assume responsibilities to reduce/limit emissions, regardless of whether or not this results in a more "significant" effort by those primarily responsible for cumulative emissions in historical terms (and therefore also, to a greater extent, for current atmospheric concentrations of GHGs).[20]

So far, the high expectations that non-Annex I countries had placed on the investments they could potentially receive through these mechanisms have not been met, and in addition the contribution that these mechanisms have made, both to the sustainable development of the host countries and to the reduction of emissions, has not been significant. Moreover, in most cases, contributions have been minimal and linked to measures and projects that have a very favorable relationship between the amount of limited/reduced emissions and the cost of doing so rather than the technology transfer involved.[21]

On the other hand, this situation introduced a significant element, from the political point of view, in the international negotiations on climate change, while it opened up the possibility of carrying out mitigation measures in countries that were not yet obliged to do so. It is clear how important the application of these mechanisms and their deepening may be for the final results that, in terms of the burden on the various socioeconomic sectors, they may have for the different countries. Mainly, because up to now the non-Annex I countries had not taken on commitments to reduce GHG emissions, not only are there growing pressures for at least the so-called "Key Developing Countries" to take on such commitments, but, since the Paris Agreement, they have already (at least) announced that they will carry out measures to limit/reduce emissions.[22] This had been clearly demonstrated in the "Copenhagen Accord" in which some countries not included in Annex I of the UNFCCC, nor in Annex B of the KP (such as China, Brazil, or Mexico, to cite only a few examples) had announced that they would be willing to assume some limitations in their anthropogenic GHG emissions, and now it has been generalized to all countries.[23]

If the main GHG emitters among the developing countries (among which Argentina may be included, depending on the indicator used) were to assume strict commitments to limit their GHG emissions, the use of the opportunities of the Kyoto Mechanisms, or the implementation of other mitigation measures (e.g., through the so-called Nationally Appropriated Mitigation Actions—NAMAs, or Low-Carbon

"secondary" markets and various national and regional systems for trading emissions titles (whether these are "certified" or merely "verified"), among which the European Union's Emission Trading Scheme (ETS), in force since the beginning of 2005, stands out. See Girardin (1998a, b, c, 2008, 2009, 2010, 2018).

[20] See CMNUCC (2015).

[21] Girardin (2009, 2010, 2018). For an idea of the true extent of the emissions involved, see www.unfccc.int.

[22] Bouille and Girardin (2002, 2003).

[23] www.unfccc.int.

Development Strategies—LCDs) could be turned against them, as long as they have exhausted the "cheapest" mitigation options to be used in these mechanisms, leaving them with the most "expensive" ones when it comes to actually having to meet their eventual reduction commitments and/or quantified emission limitation. In addition, as already mentioned, the expectations placed on the Kyoto Mechanisms, mainly the CDM, have not been met either.[24]

However, this is not the only conflict faced by countries like Argentina. There is a growing awareness that the mitigation measures proposed at the international level and that are beginning to be implemented will not be sufficient to avoid the consequences of climate change.[25] Then, some degree of adaptation to the effects of climate change will be necessary. In this sense, there is a conflict about how many resources to devote to adaptation and how many to climate change mitigation. The development and transfer of technology can play an important role, not only in mitigation, but also in adaptation to climate change. New resources will be needed to devote to these new challenges.[26] In this sense, the instruments that emerge from the "Cancun Agreements" (COP16/MOP6), subsequently deepened in Durban (COP17/MOP7), Doha (COP18/MOP8), and Warsaw (COP19/MOP9), appear to be very limited in achieving these objectives. As with so many other issues, everything is being delayed to be dealt with at the next COP in Glasgow, which does not yet have a definite date due to the COVID-19.[27]

The entry into force of the Kyoto Protocol (KP), beyond the modest commitment to reduce greenhouse gas emissions (GHG) that arise from it, implied a very important step in the international negotiation on climate change (in fact, its importance is evident if one considers how difficult it was to define what type of commitment would be assumed in the second commitment period of the KP and what magnitude these commitments would have), but this fact did not mean that the prevailing uncertainties on the subject had been cleared up. Neither has the signing of the Paris Agreement (by itself).

In this sense, it is worth remembering that until now, the main emitter, both current and historical, of GHGs at the international level (the USA), was still outside the

---

[24] Girardin (2009). See www.unfccc.int.

[25] CEPAL (2010), Stern (2006a, b), IPCC (2013b, 2014a, b), and PNUMA (2019).

[26] CEPAL (2010), Girardin (2008), and Girardin et al. (2017a, c).

[27] The main instruments arising from the Cancun Agreements are the following: Green Climate Fund (which consists of long-term financing for developing countries, to finance projects, programs, measures, and actions. This fund has a Transition Committee and is still being developed); the Technology Mechanism: (aimed at facilitating the development and transfer of technology, both for adaptation and mitigation, and is made up of an Executive Committee and a Network of Climate Change-related Technology Centers); the Cancun Adaptation Framework (with the aim of supporting adaptation activities through international cooperation, taking into account the urgencies of those countries that are most vulnerable), the Fast Start Finance (which involves the mobilization of additional funds, in the order of US $30 billion, over the period 2010–2012, balancing adaptation funds with mitigation funds), and finally, the establishment of Forest Management Reference Levels (for those countries included in Annex I). For more details, see the websites of the Forum on Climate Change and Trade (www.ambienteycomercio.org) and the UNFCCC Secretariat (www.unfccc.int).

rules of the game followed by the rest of the countries, which implied a high degree of uncertainty in the problem. As mentioned above, almost thirty years after the signing of the UNFCCC, many things have changed. One of the most important is that emissions from the country that has been the largest emitter to date have recently been surpassed by those of another major player on the international scene (China), which had not yet made quantified commitments to reduce emissions, but which is growing rapidly and incorporating large masses of rural people into urban life.[28] That is why the analysis of the effect that the Agreement that emerged from COP-21 may have (insofar as it also incorporates these two great actors, previously "absent") become essential.

The combination of attitudes toward taking on mitigation commitments by the two largest GHG emitters left more than 50% of total GHG emissions outside the control of the Kyoto Protocol (in its second commitment period). This situation was further exacerbated by the fact that countries such as the Russian Federation, Canada, and Japan had announced that they would not participate in a second commitment period under the Kyoto Protocol. The new rules that will necessarily have to be imposed and deepened as a result of the Paris Agreement will certainly have differential social, economic, and political effects for the different actors in the international negotiation on climate change. However, this heterogeneity in the expected economic impacts has not been sufficiently analyzed or made explicit in most of the studies on the subject.[29]

On the other hand, there is a growing consensus that the measures that are being implemented, and those that are announced in the short and medium term (including the NDCs that arise from the Paris Agreement), will not be enough to avoid changes in the climate that will impact on the living conditions of both humans and other species. This means that, beyond the mitigation efforts made, it will also be necessary to devote resources to measures of adaptation to climate change and climate variability. This issue is particularly important for the non-Annex I countries, whose responsibility for having reached the current situation is less, but whose vulnerability (measured under different aspects and indicators) and the consequent costs of adaptation will surely be disproportionately high in relation to their incidence in the solution of the problem.[30]

In this context, at least five themes are of primary relevance:

(a) the growing conflict that will exist between dedicating resources to adaptation (the main need of non-Annex I) or devoting them to mitigation (the main obligation of Annex I), despite the fact that today it is observed that most of the funds available by non-Annex I countries are dedicated to issues related to climate change mitigation. The funds available for adaptation are scarce, with the excuse that their benefits are local rather than global;
(b) the increasing importance that technology development and transfer can have, not only in mitigating, but also in adapting to climate change;

---

28 See www.unfccc.int and www.wri.org.

29 Girardin (2000), Girardin et al. (2014), and Lipietz (1995).

30 IPCC (2013a, 2014b), Girardin (2008, 2013), and Stern (2006a, b).

(c) beyond the announcements made in the Paris Agreement, what degree of deepening the concrete commitments to be assumed by Annex I countries for the post-2012 period (and beyond) and which non-Annex I countries will have to assume effective quantitative emission reduction commitments and, at the same time, what characteristics the commitments to be defined for them will have;
(d) how to address the adaptation needs of the most vulnerable populations and what criteria will be used to set priorities;
(e) how the consequences of COVID-19 will impact on vulnerability, adaptation, and mitigation in the short, medium, and long term.

So far, the announcements made at the latest meetings of the Conference of the Parties (COP, related to the Climate Change Convention) and the Meeting of the Parties (MOP, related to the Kyoto Protocol) and embodied in the Paris Agreement are very vague and, in some cases, very modest (such as the proposed emissions reductions announced by the USA which, by changing the base year from 1990 to 2005, does not imply a significantly greater commitment than that stipulated in the KP).[31]

Given this situation, it is unavoidable to deal with which additional instruments will be developed at the international level to continue the process begun with the UNFCCC, followed by the Kyoto Protocol and continued with the Paris Agreement. Instruments that, beyond mitigation, will not be able to ignore the need for adaptation to the effects of climate change: not those expected in the future, but those that are already being observed today, mainly in the case of non-Annex I countries.

Within this framework, the various scenarios designed to continue the process initiated by the UNFCCC, continued by the KP, and reaffirmed by the Paris Agreement involve various combinations of opportunities and risks that must be evaluated in terms of the impacts and consequences they may have on the various sectors and economic activities that could benefit and/or be harmed by the measures taken. A thorough analysis of these possibilities is of enormous importance, at the national level, to assess the potential effects on the economy of the various sectors and the country as a whole.[32]

It is also a very significant fact that the first commitment period of the Kyoto Protocol ended at the end of 2012, 20 years after the signing of the UNFCCC. In this sense, it is necessary to rethink some issues as many particular situations that determined some of the premises included in the Convention have changed during this time. The Kyoto Protocol itself proved to be insufficient, more so because some countries that had ratified it (the aforementioned Japan, Canada, and Russia, but also less important emitters such as New Zealand, among others) were not willing to sign up to a second commitment period with quantitative targets in absolute terms. It is time to analyze whether those that announced measures to reduce and/or limit emissions (USA and China, as the most prominent) comply with these announcements. It

[31] Girardin (2008, 2013), Girardin et al. (2014), CEPAL (2010); www.unfccc.int.

[32] CEPAL (2010) and Girardin et al. (2017a).

remains to be seen if, since the Paris Agreement, significant progress is indeed being made in the objective of avoiding human interference with the climate.[33]

The problem of climate change is of enormous complexity, trans-disciplinarily, and amplitude, as it involves aspects related to various sciences, from meteorology to economics. Consequently, it seems pertinent to start by precisely delimiting the objectives, within a framework of analysis that must necessarily take into account that it is a problem with a great dynamism in terms of the constant incorporation of new knowledge, tools, and empirical evidence.

Some of the hypotheses that can be raised when carrying out this analysis, from the point of view of a Latin American country, are the following:

- Current atmospheric concentrations of greenhouse gases depend not only on current but also on past emissions. In this sense, it is notorious the responsibility of the industrialized countries, which have been emitting the largest proportion of these gases since the Industrial Revolution until today. This is important not only from the point of view of mitigation, but also from that of adaptation. There is a growing need for the inhabitants of the planet to adapt not only to climate change (understood as long-term changes), but mainly to climate variability (measured in much shorter periods). Many of the countries most affected by the potential impacts of climate change are at the same time those that contributed the least to the problem and, therefore, independently of the mitigation measures that they apply, and they will also have to dedicate enormous resources to adaptation. However, delays in decision-making and in tackling the root of the problem mean that the responsibility of some of the more developed non-Annex I countries is growing. This situation will further complicate the already complicated problem of sharing the burden of mitigation costs.
- If the distribution of the burden of the costs of mitigating the effects of climate change were to be governed by equity criteria, it is undeniable that the greatest effort would have to be borne by those who contributed most to the emergence of the problem (Annex I). But, on the other hand, the non-Annex I countries will also have to have access to the financing that will allow them to carry out adaptation measures. However, most of the resources they access are committed to be dedicated to mitigation rather than adaptation.
- In turn, the International Financial Organizations and the Political and Academic Circles of the Developed Countries recommend the use of economic models and instruments that are governed by economic efficiency criteria. Thus, priority is given to where it is cheaper to mitigate (more cost-effective), regardless of the responsibility of the different actors. In fact, this logic is embedded in the mechanisms that appear with the Kyoto Protocol in order to make compliance with the commitments assumed by Annex I countries more flexible.
- Also, the results arrived at will be strongly influenced by the data and assumptions that are incorporated as inputs to the different models. If, in order to weigh the potential costs and/or benefits involved in the different mitigation measures that

[33] See Earth Negotiation Bulletin, www.iisd.ca/vol12/.

can be carried out, only indicators such as GDP are used, without taking into account the existing differences in terms of income distribution, the importance of the decisions made in the richer countries would be overestimated. Thus, the same number of jobs lost would result in a higher cost (would be valued more negatively) if it occurred in an industrialized country, simply because a higher average wage would be applied and the same would happen with other types of resources and factors of production. This situation can be further complicated by criteria such as the carbon footprint of products imported from non-Annex I countries, which if it does not take into account the entire life cycle of the product, may be placing a disproportionate burden on countries whose products suffer higher transport costs.

- In this sense, there are serious doubts as to whether the suggested criteria are indeed the most advantageous from the point of view of countries such as Argentina.
- A fundamental hypothesis of this analysis can be summarized as follows "The Legal Instruments created by the International Organizations and the documents produced by the Political and Academic Circles of the Industrialized Countries (IP) strongly recommend the application of cost-effectiveness criteria at a global level, in order to determine the optimal location of the measures that should be taken to mitigate the effects of climate change. This situation is not neutral, neither from the economic nor from the ethical point of view. The application of these principles leads to a disregard for the historical responsibility for having reached the current risk situation and does not create incentives for early action, but rather the opposite, insofar as the proposed Market Mechanisms (first in the Kyoto Protocol and later in the Paris Agreement) seem to favor large emitters and those who delayed the implementation of mitigation measures (within and outside of Annex B), over marginal emitters and those who were managed in accordance with the precautionary principle. Thus, the latter will bear disproportionately high costs in relation to their smaller relative contribution to the problem (atmospheric concentration of GHGs) and, at the same time, will not be able to devote sufficient resources to their main concern: Adaptation to the Expected Effects of Climate Change".
- In this sense, there are well-founded doubts that the agreements reached at the COP-3 in Kyoto, which were deepened throughout the international negotiation process by the COP13/MOP3 in Bali (December 2007), but fundamentally by the Summits that followed the one in Copenhagen (COP15/MOP5); the Cancun Summit (COP16/MOP6); the Durban Summit (COP17/MOP7); the Doha Summit (COP18/MOP8); the Warsaw Summit (COP19/MOP9); the Lima Summit (COP20/MOP10); and finally the Paris Summit (COP-21/MOP11) imply at least three things:
  - They are a postponement in the treatment of the problem of global climate change, in that they allow Annex I countries not to comply in the short term with the commitments they took on in ratifying the UNFCCC (stabilizing GHG emissions at 1990 levels in 2000) and the Kyoto Protocol (whose

time horizon for the first commitment period was set for 2008–2012). Subsequently, the reference date was 2020 (second commitment period of the Kyoto Protocol, from the Doha Addendum). Now, the closest time horizon to which the measures are referenced is 2030.[34] This situation transforms this era into a critical one as we have to quickly define what actions are going to be taken in the immediate and near future.

- They are a way of transferring a good part of the mitigation costs to developing countries, through the application of the mechanisms proposed to "make them more flexible" and make it possible to meet the objectives of limiting IPs' emissions, especially since the process of improving the quality of life of the population in many of these countries and the dynamics of incorporating large masses of rural people into the urban environment means that these countries are taking on increasing responsibility for climate change.
- Both situations are a way of avoiding the criterion of "common but differentiated responsibilities" explicitly mentioned in the UNFCCC and taken up by the spirit of the KP.

References

Argawall A, Narain S (1991) Global warming in a unequal world: a case of environmental colonialism. Centre for Science and Environment, New Delhi

Bhaskar V (1995) Distributive justice and the control of global warming. In: Bhaskar V, Glyn A (eds) The north, the south and the environment. Ecological constraints and the global economy. UNU, Tokyo

Bouille D, Girardin LO (2002) Learning from the Argentine voluntary commitment. In: Baumert K, Blanchard O, Llosa S, Perkaus J (eds) Building on the Kyoto Protocol. Options for protecting the climate. WRI, CSDA, CEIBA, COPPE/UFRJ, EDRC, FB, ENVORK, IFE, IPS, KEI, Washington, pp 135–156

Bouille D, Girardin LO (2003) Conditions for greater commitment of developing countries in the mitigation of climate change. In: Climate change compendium. Climate Change Knowledge Network (CCKN), IISD, Ottawa

Bouzas R (2011) Climate change mitigation and impacts on trade: challenges for Latin America. Pensam Iberoam (8), 13 Apr 2011. http://www.pensamientoiberoamericano.org/xnumeros/8/pdf/pensamientoIberoamericano-167.pdf

Criqui P, Kouvaritakis N (1997) The costs for the energy sector of reducing CO_2 emissions: an international assessment with the POLES model. Research notebook No. 13. IEPE, Grenoble University of Social Sciences

Girardin LO (1998a) Economy and climate change under the light of the Kyoto Protocol. Key issues related with the eventual implementation of the flexibility mechanisms for the fulfilment of the commitments of GHG reductions. Bariloche Foundation, Environment and Development Program, Buenos Aires, June 1998

Girardin LO (1998b) Towards the search for alternative proposals that imply concrete benefits for non-Annex I parties due to their eventual greater participation in the climate change prevention process. Bariloche Foundation, Environment and Development Program, Nov 1998

[34] Second commitment period under the Kyoto Protocol was set for 2013–2020. The closest reference period set in the Emission Gap Reports is 2030.

Girardin LO (1998c) Key issues to take into account to define the opportunities that arise for Argentina for its eventual participation in the Kyoto mechanisms, in the current international context. Study on flexibility mechanisms in the context of the United Nations Framework Convention on Climate Change (UNFCCC) and the Kyoto Protocol (KP). Government of the Argentine Republic, Government of Canada, World Bank, Buenos Aires, Dec 1998

Girardin LO (2000) Global climate change and the distribution of the mitigation costs of its possible consequences among different countries. Situation based on the results of the fifth conference of the parties (COP-5; Bonn, October 25 to November 5, 1999), Mar 2000. Master's degree in environmental and territorial policies. Institute of Geography, Faculty of Philosophy and Letters, Buenos Aires University. Director: Lic. Daniel Hugo Bouille. Approved and defended in August 2000. Qualification: Distinguished. Bariloche Foundation, Environment and Development Program (MADE). Available at www.fundacionbariloche.org.ar

Girardin LO (2008) The future of international climate change regime. Key issues to be considered. In: Leite da Silva Dias P, Costa Ribeiro W, Lima Sant'Anna Neto J, Zullo Jr J (eds) Public policy, mitigation and adaptation to climate change in South America. University of Sao Paulo, Institute on Advances Studies

Girardin LO (2009) Myths and realities of the role of the CDM, in its contribution to the fulfillment of the ultimate objective of the convention on climate change and sustainable development. Opportunities versus realities. The case of Latin America. Auton Econ (3). Latin American Autonomous University, Medellin, Colombia

Girardin LO (2010) Introduction: some socioeconomic and political aspects of climate change. In: Introduction of the "2009 annual environmental report" prepared at the request of the environmental protection agency of the government of the autonomous city of Buenos Aires, Mar 2010, 4 pp

Girardin LO (2013) Socioeconomic and political aspects of climate change. From the convention to the Kyoto Protocol. Volume I (1990–2000). Patagonia Third Millennium Foundation, Buenos Aires, Trelew, Aug 2013. ISBN 978-987-26155-8-1

Girardin LO (2018) Myths and realities of the role of the CDM and other market mechanisms in their contribution to sustainable development. Sci Res Mag 68(5):72–86. ISSN 0009-6733. Argentine Association for the Progress of Sciences (AAPC), Buenos Aires. Available at http://www.argentinaciencias.org

Girardin LO et al (coord) (2014) The economics of climate change in Argentina. First approximation. United Nations, ECLAC. Development assistance of the government of the United Kingdom. Ministry of Agriculture, Food and Environment of Spain, European Union, German Development Cooperation, IDB, Ministry of Foreign Affairs of Denmark, Santiago de Chile, Jan 2014

Girardin LO et al (coord and ed) (2017a) The economics of climate change in Argentina. Volume I: synthesis report and economic valuation report, Dec 2017. United Nations Economic Commission for Latin America and the Caribbean (ECLAC), Patagonia Third Millennium Foundation, Trelew-Buenos Aires, 366 pp. ISBN 978-987-45525-3.2. Available at http://www.patagonia3mil.com.ar/publicaciones/

Girardin LO et al (coord and ed) (2017b) The economics of climate change in Argentina. Volume II: impacts, vulnerability and adaptation, Dec 2017. United Nations Economic Commission for Latin America and the Caribbean (ECLAC), Patagonia Third Millennium Foundation, Trelew-Buenos Aires, 490 pp. ISBN 978-987-45525-4.9. Available at http://www.patagonia3mil.com.ar/publicaciones/

Girardin LO et al (coord and ed) (2017c) The economics of climate change in Argentina. Volume III: emissions and mitigation scenarios, Dec 2017. United Nations Economic Commission for Latin America and the Caribbean (ECLAC), Patagonia Third Millennium Foundation, Trelew-Buenos Aires, 564 pp and Annexes. ISBN 978-987-45525-5.6. Available at http://www.patagonia3mil.com.ar/publicaciones/

Intergovernmental Panel on Climate Change (IPCC) (2013a) Climate change. Synthesis report. In: Fifth assessment report (5AR). WMO, UNEP

Intergovernmental Panel on Climate Change (IPCC) (2013b) Climate change 2013. The physical science basis. In: Contribution of working group I to the fifth assessment report of the intergovernmental panel on climate change. Fifth assessment report (5AR). WMO, UNEP. Cambridge University Press, UK, USA, 996 pp

Intergovernmental Panel on Climate Change (IPCC) (2014a) Climate change 2014. Impacts, adaptation and vulnerability. In: Contribution of working group II to the fifth assessment report of the intergovernmental panel on climate change. Fifth assessment report (5AR). WMO, UNEP. Cambridge University Press, UK, USA, 976 pp

Intergovernmental Panel on Climate Change (IPCC) (2014b) Climate change 2014. Mitigation of climate change. In: Contribution of working group III to the fifth assessment report of the intergovernmental panel on climate change. Fifth assessment report (5AR). WMO, UNEP. Cambridge University Press, UK, USA

Le Quéré C et al (2020) Temporary reduction in daily global CO_2 emissions during the COVID-19 forced confinement. Nat Clim Change. https://doi.org/10.1038/s41558-020-0797-x

Lipietz A (1995) Enclosing the global commons: global environmental negotiations in a north-south conflictual approach. In: Bhaskar V, Glyn A (eds) The north, the south and the environment. Ecological constraints and the global economy. UNU, Tokyo

Stern N (2006a) Stern review: the economics of climate change. H. M. Treasury, London

Stern N (2006b) Stern review on the economics of climate change. Available at http://webarchive.nationalarchives.gov.uk/+/http://www.hmtreasury.gov.uk/independent_reviews/stern_review_economics_climate_change/stern_review_report.cfm

United Nations Economic Conference for Latin America and the Caribbean (ECLAC) (2010) The economics of climate change in Latin America and the Caribbean. In: Synthesis 2010, Santiago de Chile, Dec 2010

United Nations Economic Conference for Latin America and the Caribbean (ECLAC) (2020a) Latin America and the Caribbean in the face of the COVID-19 pandemic. Economic and social effects. In: COVID-19 special report No. 1, Santiago de Chile, 3 Apr 2020, 15 pp. Available at https://repositorio.cepal.org/bitstream/handle/11362/45527/5/S2000325_es.pdf

United Nations Economic Conference for Latin America and the Caribbean (ECLAC) (2020b) Dimension the effects of COVID-19 to think about reactivation. In: COVID-19 special report No. 2, Santiago de Chile, 21 Apr 2020, 21 pp. Available at https://repositorio.cepal.org/bitstream/handle/11362/45445/4/S2000286_es.pdf

United Nations Economic Conference for Latin America and the Caribbean (ECLAC) (2020c) The social challenge in times of COVID-19. In: COVID-19 special report No. 3, Santiago de Chile, 21 May 2020, 22 pp. Available at https://repositorio.cepal.org/bitstream/handle/11362/45527/5/S2000325_es.pdf

United Nations Economic Conference for Latin America and the Caribbean (ECLAC) (2020d) Compilation of national accounts, balance of payments and foreign trade statistics within the framework of the coronavirus disease (COVID-19) health emergency. In: COVID-19 reports, Santiago de Chile, June 2020, 15 pp. Available at https://www.cepal.org/es/publicaciones/45666-compilacion-estadisticas-cuentas-nacionales-balanza-pagos-comercio-exterior

United Nations Economic Conference for Latin America and the Caribbean/International Labour Organization (ECLAC/ILO) (2020) Labor situation in Latin America and the Caribbean. Work in times of pandemic. Challenges facing coronavirus disease (COVID-19), Santiago de Chile, May 2020, 60 pp. Available at https://repositorio.cepal.org/bitstream/handle/11362/45337/6/S2000264_es.pdf

United Nations Environmental Programme (UNEP) (2019) Emission gap report 2019. Available at https://www.unenvironment.org/interactive/emissions-gap-report/2019/report_es.php

United Nations Framework Convention on Climate Change (UNFCCC) (1992) United Nations Framework Convention on Climate Change. Available at https://unfccc.int/files/essential_background/background_publication_htmlpdf/application/pdf/convsp.pdf

United Nations Framework Convention on Climate Change (UNFCCC) (1997) Kyoto Protocol. Available at http://unfccc.int/resource/docs/convkp/kpspan.pdf

United Nations Framework Convention on Climate Change (UNFCCC) (2015) Paris Agreement. Available at https://unfccc.int/files/meetings/paris_nov_2015/application/pdf/paris_agreement_spanish_pdf

Web Sites

Web Site of Climate Change and Trade Forum: www.ambienteycomercio.org
Web Site of ECLAC: www.eclac.org
Web Site of IISD, Editor of Environmental Negotiation Bulletin (ENB): www.iisd.ca/vol12/
Web Site of IPCC: www.ipcc.ch
Web Site of Nature: www.nature.com/natureclimatechange
Web Site of UNFCCC: www.unfccc.int
Web Site of WRI: www.wri.org

Chapter 2
Agriculture and Climate Change: Opportunity or Threat?

Abstract This chapter is based on a presentation made at the Argentine Association of Agricultural Economics (AAEA) in which the main threats and challenges posed by climate change to agriculture, both globally and regionally, and particularly for Argentina, were discussed, taking as a source the Intergovernmental Panel on Climate Change (IPCC) assessment reports and the diverse studies carried out in Argentina and South America.

Keywords Agriculture · Climate change impacts on agriculture · Vulnerability · Adaptation · Argentina · Latin America

The climate has been in continuous change and evolution since the origins of the planet because of the diverse natural processes that influence the factors that determine it. However, the data provided by the scientific community, mainly during the 1980s, open up the possibility that current climate variations are, to some extent, driven by human action.

The starting point of this statement is a general contextualization of the climate change issue and what are the relevant aspects regarding it in the agricultural sector at a global level. This presentation will show some results of a work that was carried out mainly between 2009 and 2011 in the framework of an ECLAC project for all Latin American countries and which had a chapter for Argentina. In this chapter, one of the fundamental sectors was studied and tried to determine what the impacts and possibilities of mitigation were for the agricultural sector.

Presentation made by M.Sc. Leónidas Osvaldo Girardin at the Annual Meeting of the Argentine Association of Agricultural Economics, Buenos Aires 2014. Recorded and Edited by the Chair of General Economics—Faculty of Agronomy—University of Buenos Aires (FAUBA). Leonidas Osvaldo Girardin: Researcher of the Argentina's National Council for Research in Science and Technology (CONICET); Researcher of Fundacion Bariloche's Environment and Development Department; Professor of Economics and the Environment Area, Department of Applied Sciences and Technology, Moreno National University (UNM, Argentina).

L. O. Girardin, *Socioeconomic and Geopolitical Aspects of Global Climate Change*, The Latin American Studies Book Series, https://doi.org/10.1007/978-3-031-53246-7_2

Most of the results that will be presented come from the last IPCC Assessment Report.[1,2]

Beyond the existence of certain skeptical visions about climate change, the IPCC collects all scientific information published between one period and another. Considering this, the dominant scientific paradigm, what it shows is that the warming of the climate system is unequivocal: in most of the scientific community there is no longer any discussion about this process of warming of the climate system.

Since 1950, many of the changes that were observed were unprecedented and could not be measured or identified in the past. Both the ocean and the surface of the earth are warming. Global ice and snow volumes, with the exception of a few specific issues, are decreasing, and this trend does not seem to be reversing, but rather seems to be accelerating. The sea level has risen, and it is also a trend that climate models suggest will not necessarily stop. Greenhouse gases (GHG) concentrations are also increasing, so why are we looking so hard at GHG concentrations? Because GHG emissions are the only factor, of all those involved in climate change, over which human beings have any kind of power to manage it. Everything else depends on the entry and exit of radiation in the atmosphere, phenomena that exceed the human being activity. Human activities have a fundamental responsibility, precisely in this issue of GHG emissions. All gases have different residence times in the atmosphere and different warming potentials. In cases like carbon dioxide, which has an average residence time in the atmosphere around 100 or 150 years, the concentrations are not only related to current emissions, but also to emissions that have been accumulating for a long time.

It is common to hear that "this was the hottest year in the last decades". What this Fifth Assessment Report of Working Group 1 shows is that, over the past three decades, each of those decades has been the warmest of any decade prior to 1850. (That is, since the measurements are reasonably reliable.)

Another important issue is that the period 1983–2012 was probably the warmest 30-year period in the last 1400 years ("probably", because the IPCC also marks different degrees of confidence in the information provided). Beyond what we may pose in terms of the uncertainties or uncertainties of the future, clearly something with the climate is happening, and this can be measured.

Figure 2.1 shows how at least in these 150 or 160 years the trend is very clear. This has to do with temperature increases and what stands out is the fear of heterogeneity, which will be a crucial concept throughout the presentation.

---

[1] In 2014, the IPCC completed the Fifth Assessment Report, prepared by the three Working Groups, which comprises three contributions on physical bases; impacts, adaptation and vulnerability, and climate change mitigation, plus a Synthesis Report. The contribution from Working Group I was accepted and approved in September 2013. The contributions from Working Groups II and III were accepted and approved in March and April 2014, respectively, and the Synthesis Report was approved and adopted in November 2014. http://www.ipcc.ch/home_languages_main_spanish.shtml.

[2] The Working Group 1 is dedicated primarily to the physics of the atmosphere. Group 2 is more concerned with issues related to impacts and vulnerability, and Group 3 is concerned with economics and policy. This is the group that works on mitigation and where science is mixed with politics.

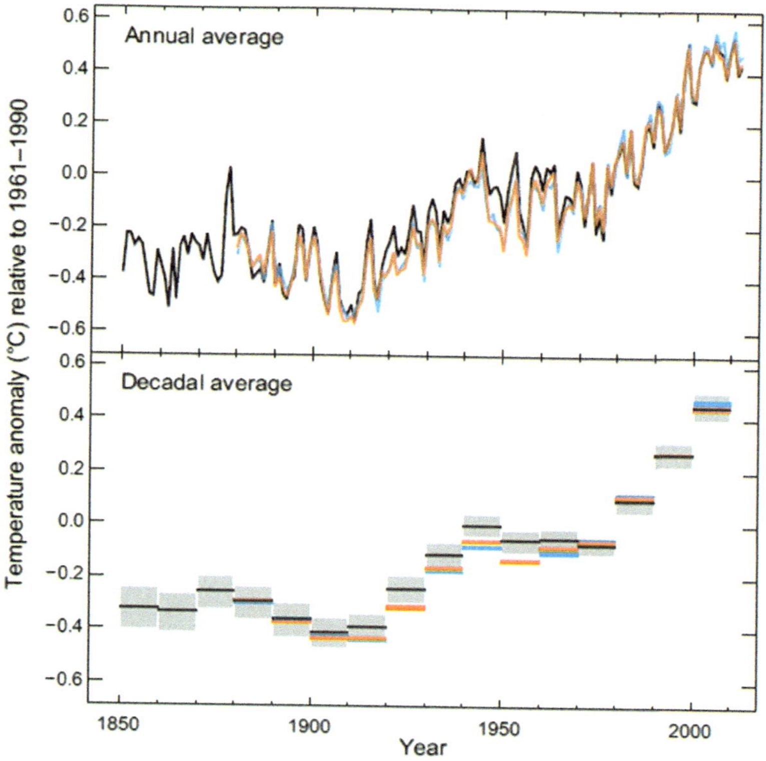

Fig. 2.1 Observed globally averaged combined land and ocean surface temperature anomaly 1850–2012. *Source* IPCC Fifth Assessment Report. WGI. SPM (2013)

It is also true that if the temperature increases much more than these 2.5° on average, which in some places has already increased, such as the Amazon, or the Tundra area in Russia (Fig. 2.2) we are at some risk, because the damage will depend on how quickly the different systems, the different species, and the different activities adjust. If this change accelerates, it deepens; obviously we are in a problem.

But if one thing is clear about climate change, it is that it does not impact everywhere equally. This does not mean that there are going to be winners and losers (because with the levels of uncertainty that exist, in the long term we will most likely all lose), but these levels of heterogeneity mean that some countries will be able to delay their decisions, and others will be exposed in such a way that they will not have any room for maneuver to delay any situation. There, temperature ranges (from the past, because these are changes observed between the beginning of the twentieth century and 2012) already mark us with some degree of heterogeneity, which we can also see in terms of rainfall (Fig. 2.3).

The brown and dark brown represent the falls in the precipitations, and the colors celestial and blue represent the increases in the precipitations. There you can see that

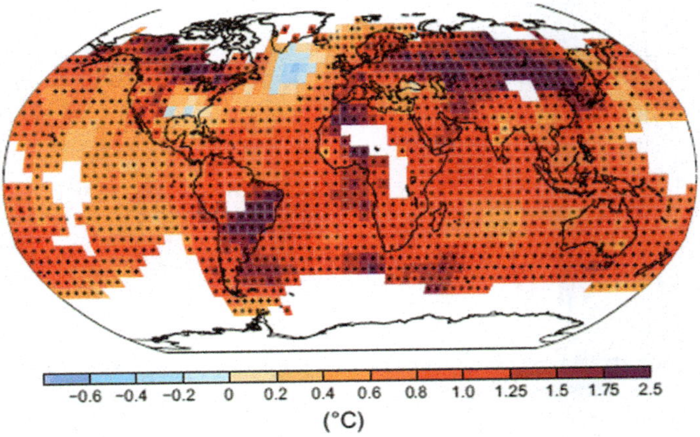

Fig. 2.2 Observed changes in surface temperatures 1901–2012. *Source* IPCC Fifth Assessment Report. WGI. SPM (2013)

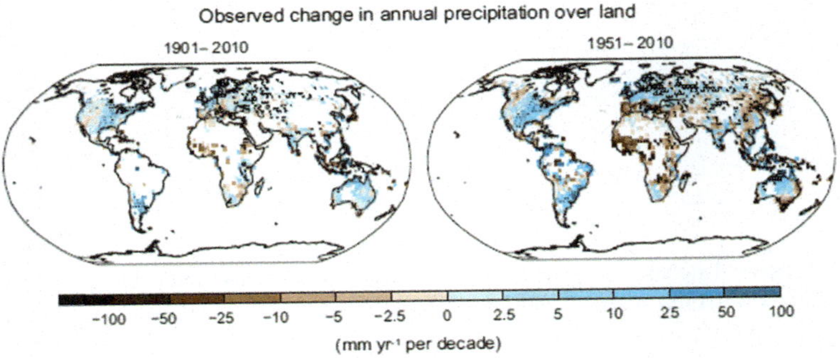

Fig. 2.3 Observed changes in annual rainfall on the ground. *Source* IPCC Fifth Assessment Report. WGI. SPM (2013)

not only those that were bad are worse off, because the whole brown area coincides with many of the desert areas, but we also see the very interior heterogeneity of our own territory.

You can see how the Humid Pampas, the Chaco area, and the Argentinean River Coast show increases in rainfall, moderate but in some cases larger, and the entire mountain range is in brown, which implies that we are already seeing drops in rainfall. Weather models do not tell us that this trend will be reversed, but quite the contrary, it seems that it will be deepened: the areas where it is raining more will continue to rain more, and the areas where it is raining less may even present some level of water stress. Also related to heterogeneity, on the left side of the film you see the whole course of this long cycle, from 1901 to 2010, and what you have on the right side is the last 60 years. There is a deepening of these changes, and they are more pronounced, either downward or upward.

Some issues that will determine the expected impacts are the issue of emissions and what will happen in the future. Everything indicates that the times of permanence in the atmosphere of the different emissions make all these phenomena have a degree of phasing or inertia, which in terms of time is very large. All those situations that are occurring, unless the changes are very significant (and when I say very significant I mean significant reductions in emissions, as to have an effect on concentrations before 2100), will be deepening. I recommend that you look at a UNEP study entitled "Emissions Gap",[3] which shows what the concentrations would be in 2020 under a scenario without much change. What it shows is that in order to prevent a temperature increase of more than 2 °C compared to pre-industrial temperatures by 2100, the current emission level has to be reduced by 60% by 2020. This important effort has a high economic cost and implies an important sacrifice in many activities. There is a major difficulty to reach this sacrifice, no matter how willing. This means that we must be prepared for many of the changes that are being predicted to take place. What it shows is that in order to prevent a temperature increase of more than 2 °C compared to pre-industrial temperatures by 2100, the current emission level has to be reduced by 60% by 2020. This important effort has a high economic cost and implies an important sacrifice in many activities. There is a major difficulty to reach, no matter how willing, this sacrifice. This means that we must be prepared for many of the changes that are being predicted to occur. Our capacity to adapt will be important, but perhaps even if we do everything possible in terms of adaptation and mitigation, we may still feel some of these impacts.

On the other hand, there are different scenarios of representative GHG concentration trajectories. Figure 2.4 shows, on the one hand, the RCP 2.6 would be one of the scenarios in which there are greater efforts to prevent emissions from increasing, and on the other hand, the RCP of 8.5 would be a case quite similar to the level of emissions mitigation we have at present.

This 2 °C limit will be reached in any of the scenarios, even in the lowest concentration scenario, that is, the one that implies a much greater effort to reduce emissions and therefore a much more costly and difficult to achieve scenario. Another thing that is also important to consider is that, except for that scenario where there is a lot of effort, for all the others, temperatures may even raise after 2100. Even for those scenarios a little higher in terms of concentrations than 2.6, such as 4.5, which is also a scenario with important efforts, this does not imply that temperatures will not continue to increase after 2100, obviously with the great heterogeneity in terms of the different regions that we saw before. This is what is proposed in terms of projection for that scenario of much greater effort and lower levels of concentration, where also the temperature increases between 0.5 and 3 °C in some cases, as in the Arctic zone, which gives us the RCP 8.5, in some cases gives us increases of up to 11 °C in the polar areas, but that gives an average increase of between 2 and 5° for most places with continental mass.

---

[3] https://www.unenvironment.org/resources/emissions-gap-report.

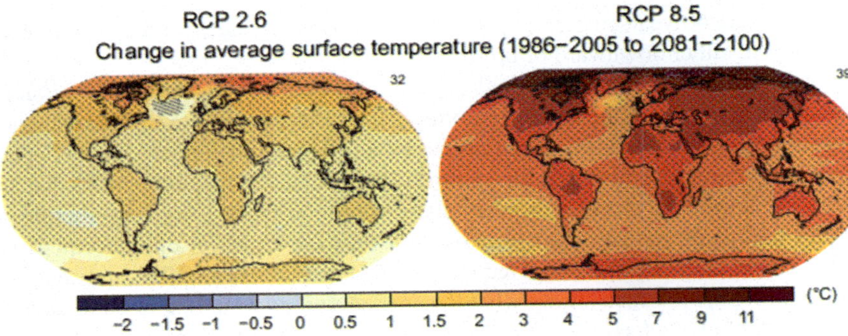

Fig. 2.4 Changes in average medium temperature on surface (1986–2005 to 2081–2100). *Source* IPCC Fifth Assessment Report. WGI. SPM (2013)

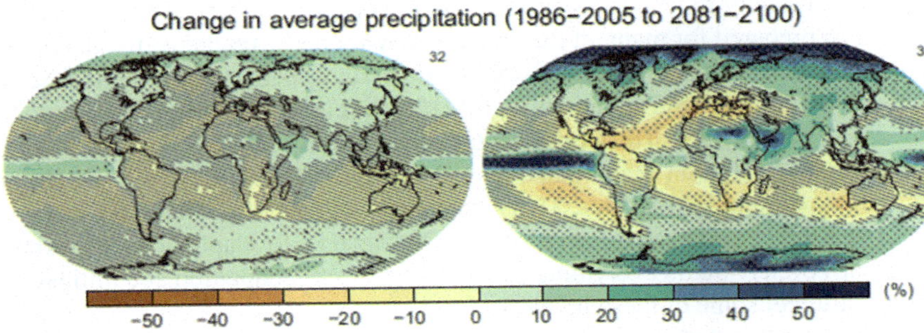

Fig. 2.5 Changes in average precipitation (1986–2005 to 2081–2100). *Source* IPCC Fifth Assessment Report. WGI. SPM (2013)

This is a rigorous scenario in terms of temperature increases, but it is not an impossible scenario: it is a scenario that is very close to the current trajectory if no measures are taken in the 80 years until 2100.

What you see in Fig. 2.5 has to do with rainfall and those two concentration paths. These are changes in average precipitation for the 4.5 scenario.

What is reflected in Fig. 2.5 is also the greater speed with which these changes occur and the greater depth with which these changes occur between 2081 and 2100. There is a very significant increase in precipitation throughout the Equatorial and Polar Regions and significant decreases in what would be the Amazon Basin and the South Pacific. There is a tendency to increase between 10 and 20% in the Humid Pampas and the Chaco Region.

With regard to the agricultural issue, it is important to note that most of these issues related to climate change will last for many centuries, even if a major effort is made to reduce GHG emissions. This, from the economic point of view, implies that adaptation measures and investments in terms of adaptation will be necessary anyway, even if we apply all the mitigation measures. We are going to have to carry

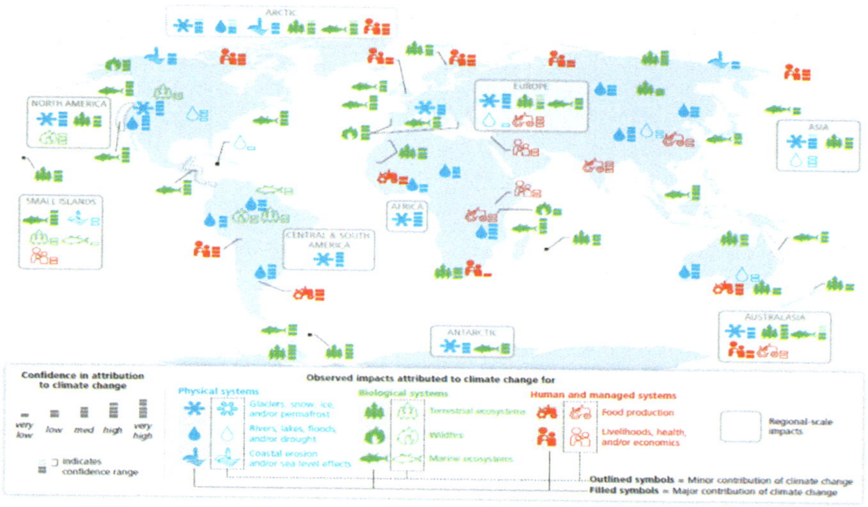

Fig. 2.6 IPCC Working Group 2 regional impacts. *Source* IPCC Fifth Assessment Report. WGII. SPM (2014)

out some kind of adaptation, because we are going to feel some kind of impact, regardless of the trajectories that the emissions show, fundamentally because of this issue of inertia that is evident.

Figure 2.6 shows that in South America, in light blue, these physical systems have to do with what will happen to the glaciers, water resources, what will happen to the coastal area, and it is a phenomenon that throughout the region, at least in South America, is very important. There is also an important issue in terms of food production, and the impacts on biological systems, mainly on land ecosystems, and in the South Atlantic with everything related to marine ecosystems; and in the Amazon region, for example, an increase in certain dangers of forest fires.

Some of the impacts for South and Central America, in terms of the impacts on water availability on semi-arid ecosystems, food production, and food quality, are shown in Fig. 2.7. There are disparities within the region that are also quite important.

Something that is important to keep in mind is the spread of vector-borne diseases. The climate systems are moving toward higher latitudes, more heat and more humidity are advancing, and this implies that in the study for Argentina, the dengue vector will come to occupy a territory in which 89% of the country's population lives by 2100. It will be necessary to be more attentive to these diseases, not only in terms of human health, but also in phytosanitary terms.

But what are the main impacts expected? It is necessary to analyze cases of snow accumulation, water resources, land ecosystems, and coastal erosion, mainly due to the rise in sea level, but also in many cases due to changes in the prevailing winds; for example, the case of the Rio de la Plata estuary in the southeastern states. Also crucial at this point is the problem of food production and human settlements.

Central and South America

Key risk	Adaptation issues & prospects	Climatic drivers	Timeframe	Risk & potential for adaptation
Water availability in semi-arid and glacier-melt-dependent regions and Central America; flooding and landslides in urban and rural areas due to extreme precipitation (*high confidence*) [27.3]	• Integrated water resource management • Urban and rural flood management (including infrastructure), early warning systems, better weather and runoff forecasts, and infectious disease control		Present Near term (2030–2040) Long term 2°C (2080–2100) 4°C	Very low — Medium — Very high
Decreased food production and food quality (*medium confidence*) [27.3]	• Development of new crop varieties more adapted to climate change (temperature and drought) • Offsetting of human and animal health impacts of reduced food quality • Offsetting of economic impacts of land-use change • Strengthening traditional indigenous knowledge systems and practices		Present Near term (2030–2040) Long term 2°C (2080–2100) 4°C	Very low — Medium — Very high
Spread of vector-borne diseases in altitude and latitude (*high confidence*) [27.3]	• Development of early warning systems for disease control and mitigation based on climatic and other relevant inputs. Many factors augment vulnerability. • Establishing programs to extend basic public health services		Present Near term (2030–2040) Long term 2°C (2080–2100) 4°C	Very low — Medium — Very high Long term: not available / not available

Central and South America

Snow & Ice, Rivers & Lakes, Floods & Drought	• Shrinkage of Andean glaciers (*high confidence*, major contribution from climate change) • Changes in extreme flows in Amazon River (*medium confidence*, major contribution from climate change) • Changing discharge patterns in rivers in the western Andes (*medium confidence*, major contribution from climate change) • Increased streamflow in sub-basins of the La Plata River, beyond increase due to land-use change (*high confidence*, major contribution from climate change) [27.3, Tables 18-5, 18-6, and 27-3; WGI AR5 4.3]
Terrestrial Ecosystems	• Increased tree mortality and forest fire in the Amazon (*low confidence*, minor contribution from climate change) • Rainforest degradation and recession in the Amazon, beyond reference trends in deforestation and land degradation (*low confidence*, minor contribution from climate change) [4.3, 18.3, 27.2-3, Table 18-7]
Coastal Erosion & Marine Ecosystems	• Increased coral bleaching in western Caribbean, beyond effects from pollution and physical disturbance (*high confidence*, major contribution from climate change) • Mangrove degradation on north coast of South America, beyond degradation due to pollution and land use (*low confidence*, minor contribution from climate change) [27.3, Table 18-8]
Food Production & Livelihoods	• More vulnerable livelihood trajectories for indigenous Aymara farmers in Bolivia due to water shortage, beyond effects of increasing social and economic stress (*medium confidence*, major contribution from climate change) • Increase in agricultural yields and expansion of agricultural areas in southeastern South America, beyond increase due to improved technology (*medium confidence*, major contribution from climate change) [13.1, 27.3, Table 18-9]

Fig. 2.7 Impacts on water availability regarding semi-arid ecosystems in Central and South America. *Source* IPCC Fifth Assessment Report. WGII. SPM (2014)

The expected impacts of climate change could imply an increase in average temperatures between 2.6 and 4.8 °C by 2100, given the range of possibilities that we had, depending on which concentration trajectory scenario we took to make the estimate.

In some regions the impacts of climate change are already reducing crop yields of some species, and if temperatures continue to raise nothing suggest that this trend will not continue. All this, combined with increases in food demand, which depends on how much the temperature rises, could lead to situations where food security is put at risk.

Globally, the crops that are identified as the most important are wheat, rice, and corn. From the point of view of the contribution of the agricultural sector, in global terms, the emissions of the agricultural sector (because agriculture includes livestock, because a large part of these emissions correspond to livestock), the weight of the sector within the total emissions is between 10 and 12% of total emissions from human activities in 2010. Its importance is relevant, but at the same time the emphasis in recent years on emissions from the agricultural sector, and primarily from livestock, is down. Energy still accounts for more than 75% of total emissions.

Some countries often argue around the problem that energy emissions are comfort emissions, and agriculture emissions are subsistence emissions, because they are for creating food. This discussion is not entirely settled. What is important is that this also suggests that there is some kind of margin within the sector, such as trying to limit or reduce emissions, which in some cases also means increasing productivity; this, within the threats or challenges, is also an opportunity. Many of these mitigation options (which do not necessarily involve reducing emissions in absolute terms, but may involve increasing productivity, and for example producing the same thing by emitting less, or emitting the same thing by producing much more in terms of kilograms of meat or liters of milk or quintals of grain) imply the possibility of applying certain measures that may lead to an increase, not only in economic performance, but also in the contribution in terms of limiting emissions. Those gases in which the agricultural sector is clearly the main contributor are methane and nitrous oxide. Methane, not so much from rice fields, but from enteric fermentation of livestock and nitrous oxide, mainly from agricultural land use, both through direct and indirect nitrogen fixation by legumes, and through the imposition of fertilizers.

Just as we saw the impacts in terms of food security, the IPCC report also states that if the temperature increases by 3° or more, agriculture's capacity to adapt may be exceeded, especially in those regions that are at low (intertropical) latitudes.

We had already mentioned the potential of the agricultural sector within the possibilities of implementing measures to limit or reduce GHG emissions. Some of these opportunities are related to the reduction or greater efficiency in issues related to emissions in land use. Direct seeding, for example, implied a reduction of at least those indirect emissions, since by using less diesel oil in the lower use of agricultural machinery; it indirectly contributes to the reduction of emissions. Soil management, livestock management, and enteric fermentation emissions depend primarily on the digestibility of the feed. Often, with changes in the animals' diet, higher yields are generated in terms of meat and milk, and GHG emissions per unit weight are also reduced. Another issue is carbon capture and storage in soils and biomass. The forestry sector has long been an issue, but it is also happening in permanent crops and annual pastures.

Biofuels (looking at the entire life cycle of that product) in some sense also have a role to play in terms of GHG emissions. In the case of Brazil, bioethanol is a good example of an energy balance. But not everywhere the energy balance of the energy spent to make biofuels is offset by what is later obtained as a final result, in energy terms.

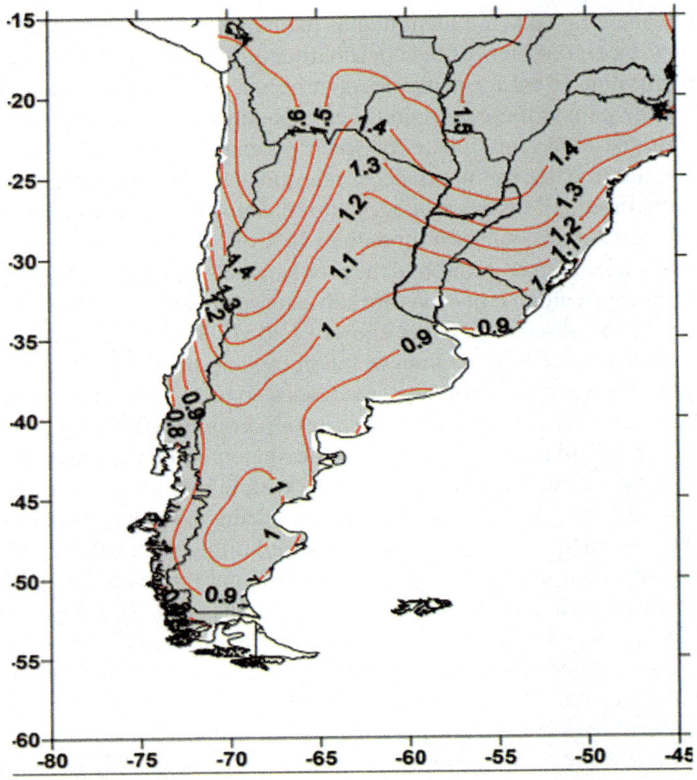

Fig. 2.8 Change in average annual temperature (°C) for the period 2020–40 with respect to 1961–90. *Source* Argentina 2CN (2006)

Climate change, broadening the issues related to heterogeneity, also generates expectations of greater volatility in the prices of agricultural products, since the average temperature is expected to increase (but 15 °C of average temperature can be between 14 and 16 °C or between 0 and 30 °C) (Fig. 2.8).

Another problem is the increase in frequency and magnitude of extreme events, which are not only related to tropical storms and cyclones, but also to extreme temperatures. The last decade had many years with higher average temperatures than those recorded in Argentina, and yet in 2007 it snowed in Buenos Aires. Nothing guarantees that this increase in the average temperature will not disguise some extremes that statistically, in these very long-term climate models, remain hidden.

An additional issue is related to the possibility of reducing food quality, in terms of different factors: related to water scarcity, with the fact that in some cases it is necessary to use more resistant species and not necessarily those of higher quality.

In general terms, it is expected at the global level (because this depends on each region and in other regions the results are not necessarily the same in the short and medium term) that the yields of the main crops will fall, that this will have some impact on the food security situation, and that these impacts will occur in a context in which there will be an increase in the demand for food. This demand will be due not only to the increase in population, which will be much lower than expected 20 years ago due to the development processes and urbanization (which causes the birth rate to fall naturally and families to have fewer children), but also because many portions of the population of several emerging countries (which still have very low living standards) are improving their incomes, and this is causing an increase in the general demand, including for food.

As for the study mentioned at the beginning, it can be said that although work began in 2008 (it was done between 2008 and 2011) it was not published until early 2014. This study has as a background the Stern Report, which was made at the request of the Ministry of Finance of Great Britain to be discussed at the COP15 in Copenhagen in 2009, in the sense of what were the real costs of not acting in terms of climate change. What the Stern Report wanted to prove was that for developed countries (with a high share of total world emissions) not taking mitigation measures meant, in terms of expected impacts, a much higher cost than implementing such measures.

This study was intended to be replicated for Latin America. If we take into account all of Latin America, such as Brazil or Mexico, it barely reaches 5% of total global emissions. Therefore, the relevance of not acting has high expected impacts, and it is important to know them to try to reverse them somehow. All these countries had already carried out a series of studies, in physical terms at least. In both the first and second National Communication on Climate Change in the case of Argentina and in some other studies of countries like Mexico, which are in the OECD, there is already a number of impacts identified, but still need to know more precisely the economic costs to address them. If we have to put a price on how soybean, wheat, and corn crops will vary, there is an international price, and the tons of increase in that yield are multiplied by the international price. Then there will be a discussion about what discount rate we are going to use; we have to use many discount rates, as we did indeed.

However, it is much more difficult to reach a consensus and find a value that everyone considers appropriate for the impact on an ecosystem or the impact on a species, than on a phenomenon such as flooding in a place where there are human settlements, taking into account that the value of the property is in those settlements. But what happens if a historical monument is flooded, what value is placed on it? Then there is a difficulty, which in some cases may be insurmountable.

This exercise served as a starting point, from which the information was collected and systematized. This presented a series of disadvantages, because there are meteorological data from 1880 until the railroads were privatized; from 1994 onward, there are places where there are no records. In these works, the main result is to know what is missing, and what is needed to make these studies much more accurate, much more reliable, and reduce as much as possible the degree of uncertainty.

Something that is not included in this work and is very important, especially in the agricultural sector, is the impact of adaptation or mitigation measures taken by third countries, such as the carbon footprint. It could not be considered because the necessary information is not available, but it is a topic that is being studied in the Third National Communication on Climate Change. It is necessary to take into account that our products come from a country that is very far from international markets: we have to see what this carbon footprint measures, because if it does not measure the entire life cycle of the product but only the cost (the emissions in terms of carbon dioxide from transport), until our products go to international markets they are obviously going to have a carbon load that is going to be impressive compared to the rest. This discussion took place between New Zealanders and the English, when the latter proposed to put a tax on the carbon emissions of sheep meat coming from New Zealand. The New Zealanders have 70 million sheep; the export of sheep meat is an important item within their trade balance. What the New Zealanders proposed was: let us make a carbon footprint, but let us take the whole life cycle of the product. The English, what do they feed their sheep? Balanced feed (they feed it oil). New Zealanders feed grass, so let us take stock of everything. Those impacts, and the potential application of border taxes, for carbon content were not taken into account, but it is another scale of changes that could occur.

Some of the sectors that were taken into account with the work team were: the hydrological sector, in terms of various basins; on agriculture, the core zone and the northwest of Argentina; and in Iberá and deforestation in the NOA; health, of the two diseases of which we had models (in which we had to load was the temperature and the expected precipitation to know which was the vector that moved). In terms of mitigation, the different emission sectors that are included in the study are those that enter in the inventory of greenhouse gases. Obviously, in the case of all emitting sectors, we must have a basic socioeconomic scenario that tells us how this activity will develop, because how many emissions we will have to wait or not depends on this development.

It is a challenge to continue updating these studies because at one time we agreed that there were few viable options to generate energy on a massive scale in order to supply the demand and, among them, nuclear energy stood out, but this is a disproportionate amount if the high figure of 20% of nuclear energy is used to satisfy the energy demand, but no other options appeared to us that were economically and technically viable to satisfy it.

A pending task in terms of updating this study is to adapt it (depending on how international negotiations are progressing) to the objectives and goals set by the different countries for 2020/2030.

In another study similar to that carried out in Uruguay (Table 2.1) for example, agriculture is responsible for almost 80% of gross emissions, not net emissions, since the forestry sector absorbs in net terms, many of the carbon dioxide emissions, but as you can see the agricultural sector is almost five times larger than the energy sector.

These are some of the impacts that arise, trying to regionalize a little, for example in Uruguay are expected increased rainfall, increased temperatures, which do not come to neutralize this increase in rainfall, and also a phenomenon similar to that experienced by a part of Argentina, in terms of the southward shift of vectors of certain diseases (Table 2.2).

Table 2.1 Uruguay: GHG emissions by category 1990–2002 (Gg of CO_2e)

Categories	1990	1994	1998	2000	2002
Energy	3641	3970	5436	5179	4107
Industrial process	230	279	518	392	253
Agriculture	21,424	22,897	23,176	21,092	22,694
Land use, land use change, and forestry	– 3047	– 6336	– 7270	– 14,210	– 23,474
Waste	1155	1228	1332	1426	1406
Totals	23,404	22,099	23,292	13,880	4986

Source Adaptation from "La economía del cambio climático en el Uruguay". ECLAC (2011)

Table 2.2 Uruguay: impacts by sector

Sector	Expected impacts	Economic valuation assumptions
Agriculture	Change in crop productivity, livestock productivity, and forestry	Production prices valued at probable long-term market prices determine sector aggregate values
Energy	Increased demand and changes in supply, addressed by the inclusion of conventional thermal sources	The new energy needs are assessed on the basis of the expected oil price
Tourism	More sun and beach tourists due to the increase in temperature, but fewer tourists due to erosion and flooding	Expenditure per additional tourist was estimated
Drinking water	Changes in the demand for drinking water	The marginal cost of drinking water consumption was applied to the calculated changes
Coastal resources	Destruction of homes and infrastructure, flooding of land, flooding, and erosion of beaches, affecting tourism	Assignment of market prices and values to each lost or affected item
Biodiversity	Changes in the products generated in Uruguay by terrestrial ecosystem services	Use of different economic valuation criteria for these products at the international level, except in the case of wetlands, which were assigned a local valuation
Extreme hydrometeorological events	Affecting the population's income, agricultural production and its industrial and commercial chains, housing, horticultural equipment, company assets, transfers, accommodation, food, infrastructure, and health care, among other areas	Estimation of lost wages, cost of home repair or construction, equipment prices, economic losses of assets and lost profits of businesses, cost of flood care, cost of infrastructure and repairs, and payments of exemptions and subsidies, among others

Source Adaptation from "La economía del cambio climático en el Uruguay". ECLAC (2011)

Table 2.3 Total cumulative costs of climate change impacts to 2050

Percentages of 2008 GDP accumulated to 2050									
Sectors	Annual discount rate 0.5%			Annual discount rate 2.0%			Annual discount rate 4.0%		
	A2	B2	Average of both scenarios	A2	B2	Average of both scenarios	A2	B2	Average of both scenarios
Agriculture	– 9.3	– 15.7	– 12.5	– 7.5	– 10.6	– 9.0	– 5.6	– 6.6	– 6.1
Energy	16.0	10.2	13.1	10.4	6.6	8.5	6.2	3.9	5.0
Tourism	– 3.6	– 2.5	– 3.1	– 2.6	– 1.9	– 2.3	– 1.8	– 1.4	– 1.6
Water	0.9	0.7	0.8	0.6	0.5	0.6	0.4	0.3	0.4
Coastal resources	2.1	0.9	1.5	1.2	0.5	0.8	0.6	0.2	0.4
Biodiversity	14.2	7.4	10.8	10.3	4.0	7.1	7.2	1.7	4.5
Disasters	10.7	3.0	6.9	7.5	2.0	4.8	5.0	1.2	3.1
Subtotal	31.0	4.0	17.5	20.0	1.0	10.5	11.9	– 0.6	5.6
Indirect	31.2	1.7	16.5	20.1	– 0.3	9.9	11.6	– 3.1	4.3
Total	62.2	5.8	34.0	40.1	0.6	20.4	23.5	– 3.7	9.9

Source Adaptation from "La economía del cambio climático en el Uruguay". ECLAC (2011)

With respect to impacts in terms of GDP, in the case of Uruguay, the 2008 impact was used. This has the impact of the agricultural sector, which is negative here, meaning that there are gains, yields increase, and costs are negative, since what is being considered here are the total accumulated costs (Table 2.3).

What we do see is that the expected yield increases in Uruguay do not have the impact of the expected yield increases in Argentina.

The two fundamental concepts are, on the one hand, heterogeneity and, on the other, vulnerability. The important thing about the idea of vulnerability is that it does not depend only on the magnitude of the impact. Vulnerability depends fundamentally on the capacity to react, on the capacity to adapt to these changes, and the latter does not necessarily depend only on the availability of economic resources, but also on other issues, and this is important to take into account.

The climate is changing; the temperature is increasing as well as rainfall and carbon dioxide (i.e., greenhouse gases are increasing). We are having this problem to which no one is indifferent: we all realize that the climate is changing, of that there is no doubt. What are the possibilities of implementing measures? So far there have been no concrete measures at the governmental or institutional level; and this is not only happening in Argentina, but in almost the whole world.

References

Argentine Republic Government (2006) Second national communication to the United Nations Framework Convention on Climate Change, Buenos Aires, 198 pp

Intergovernmental Panel on Climate Change-IPCC (2013) Climate change 2013: the physical science basis. In: The Working Group I contribution to the fifth assessment report of the Intergovernmental Panel on Climate Change (IPCC). Summary for policy makers. Cambridge University Press, 28 pp

Intergovernmental Panel on Climate Change-IPCC (2014) Climate change 2014: impacts, adaptation and vulnerability. In: The Working Group II contribution to the fifth assessment report of the Intergovernmental Panel on Climate Change (IPCC). Summary for policy makers. Cambridge University Press, 34 pp

United Nations Economic Commission for Latin America and the Caribbean-ECLAC (2011) The economics of climate change in Uruguay. Santiago del Chile, 77 pp

Chapter 3
Climate Change and Semi-arid Regions in Latin America Threats and Challenges

Abstract This chapter refers to the threats and challenges of semi-arid regions in Latin America related to climate change. According to the definition of the Intergovernmental Panel on Climate Change, the vulnerability of a system to the adverse effects of both climate variability and change is a multidimensional phenomenon, which is a function of the exposure to the phenomenon, the sensitivity to it, and the adaptive capacity (or resilience) that the system possesses. It is expected that in the near future, Latin America and the Caribbean will face serious consequences as a result of climate change. These effects will not be evenly distributed and some semi-arid areas of the region will be particularly affected. This study specifically addresses four key areas: the Gran Chaco in Argentina, Bolivia, and Paraguay; the Northeast of Brazil; the Venezuelan Coast; and the so-called Central American Semi-arid Belt. These regions are very important for the countries involved, mainly because of their role in agricultural production and consequently in food security and economic activity in their respective societies; hence the importance of analyzing the potential impacts and possible adaptation measures that can be developed.

Keywords Semi-arid regions in Latin America · Climate change impacts on vulnerable ecosystems · Adaptation · Vulnerability

3.1 Introduction

Vulnerability to climate change is a multidimensional phenomenon. According to the Intergovernmental Panel on Climate Change (IPCC), vulnerability can be defined as the degree to which a system is susceptible to the adverse effects of climate change (including climate variability and extremes) or the extent to which it is unable to cope with them. Vulnerability to climate change will be defined as a function of exposure,

Leónidas Osvaldo Girardin. National Council for Research in Science and Technology of Argentina (CONICET)/Bariloche Foundation.

L. O. Girardin, *Socioeconomic and Geopolitical Aspects of Global Climate Change*, The Latin American Studies Book Series, https://doi.org/10.1007/978-3-031-53246-7_3

sensitivity, and adaptive (or resilient) capacity (IPCC 2007, 2014b). Thus[1]:

$$\text{Vulnerability} = (\text{sensitivity} + \text{exposure}) - \text{adaptive capacity} \tag{3.1}$$

This concept of vulnerability to climate change becomes very important for the purpose of using it to assess the potential nature, extent, and severity of climate change impacts in various places, while enhancing adaptation efforts. In this regard, according to a recent report prepared by the Andean Development Corporation/ Latin American Development Bank (CAF), in the near future the Latin American and Caribbean region (LAC) will face serious consequences as a result of climate change. These effects will not be distributed evenly. This study presents an analysis of the region's degree of vulnerability, measured by the Climate Change Vulnerability Index (CCVI), which evaluates the different countries according to the risk of exposure to climate change and extreme events in terms of human sensitivity to these changes and the capacity to adapt. In this ranking, Guatemala appears in 2nd place, El Salvador in 3rd place, Honduras in 4th place, Nicaragua in 6th place, Paraguay in 8th place, Bolivia in 9th place (all of them categorized as extreme risk), the Bolivarian Republic of Venezuela is in 11th place (high risk), while Brazil and Argentina are categorized as medium risk, among the 33 countries analyzed by the study (CAF 2014).

This vulnerability will depend not only on issues related to climate change (longer term) but also on climate variability (corresponding to shorter periods). In this regard, the temperature increase scenarios for the region show projections of a broad warming pattern ranging from 1 to 2 °C by the middle of the twenty-first century (IPCC 2013). In fact, over the last three decades the entire Latin American region (except the coasts of Chile and Peru) has experienced an average increase ranging from 0.3 to 1.5 °C, especially the northern regions of South America (CAF 2014). But, beyond the impacts expected in longer terms, climate variability also influences. The main factor in the current year-to-year variability of climate change in the LAC region is the El Niño-Southern Oscillation (ENSO) phenomenon, which plays an important role in the considerable temporal and spatial variation of climate-related extreme events. ENSO is generally associated with drier than normal conditions in Central America, the Caribbean, and Mesoamerica, while La Niña is associated with conditions of higher rainfall. With respect to South America, El Niño is attributed to drought conditions in the northeast of Brazil, Venezuela, and Colombia, while it

---

[1] According to the IPCC definition (2014c): **Vulnerability** is the degree to which a system is susceptible to, or unable to cope with, adverse effects of climate change, including climate variability and extremes; exposure is the character and degree to which a system is exposed to significant climate variation. **Sensitivity**: the degree to which a system is affected—adversely or beneficially—by climate-related stimuli. **Adaptive Capacity** is the ability of a system to adjust to climate change (including climate change variability or variations) in order to achieve moderate potential damage, take advantage of opportunities, or cope with consequences. It is also defined as the set of capacities, resources, and institutions of a country or region to implement effective adaptation measures. **Adaptation**: the ability of natural or human systems to accommodate themselves in response to actual or expected climatic stimuli or their effects. This capacity to accommodate cushions the damage or takes advantage of beneficial opportunities.

Table 3.1 Classification of aridity conditions

Aridity regime	Conditions
Xeric	12 dry months and $I < 0.005$
Hyper-arid	11–12 dry months
Arid	9–10 dry months
Semi-arid	7–8 dry months
Subhumid	5–6 dry months
Humid	3–4 dry months
Hyper-humid	1–3 dry months
Hydric	0 dry months and $P < 2500$ mm
Hyper-hydric	0 dry months and $P > 2500$ mm

Source UNESCO/PHI-LAC (2010)[2]

is attributed to stimulating the appearance of floods along the western coast of the continent, mainly in the southeast (IPCC 2013; CAF 2014; World Bank 2014).

Understanding the particular components that affect vulnerability in each country makes it possible to clarify in which sectors and from which actions efforts aimed at building resilience can be directed. In this regard, it is particularly important to recognize that, in this context, the challenges posed by climate change represent global issues that require a comprehensive approach to address them. However, the impacts of climate change will not be uniform across the LAC region, among other things, because of varying degrees of vulnerability. The magnitude and severity of the impacts of climate change are subordinated to the climatic, topographical, socioeconomic, and political factors inherent in the region and are constituted in place-specifically. For example, some parts of Central America are subject to major impacts from droughts, cyclones, and the El Niño-Southern Oscillation (ENSO) phenomenon. Considering the importance of agriculture for the national economies of Central American countries, the impacts of climate change are likely to increase (CAF 2014).

These potential impacts (mentioned above) will overlap with other existing conditions that mark different situations of vulnerability to climate change in some sectors, such as agriculture, that are highly dependent on both prevailing climatic conditions and other factors. In this sense, Table 3.1 and Map 3.1 show a classification of the territory of Mesoamerica, Central America and South America, based on their levels of aridity (UNESCO/PHI-LAC 2010). There, it can be seen how in large extensions of this large region, conditions ranging from Xeric (with 12 dry months) to subhumid (with 5–6 dry months) prevail. Some of these regions (Northeast Brazil, Coastal Zone of Venezuela, Dry Corridor of Central America and the Gran Chaco American area) will be studied in this document.

---

[2] I = Aridity Index (annual precipitation/annual reference evapotranspiration). An index lower than 0.05 is considered hyper-arid. $P < 2500$ mm and $P > 2500$ mm are precipitation greater and less than 250 mm per year, respectively (UNESCO/PHI-LAC 2010).

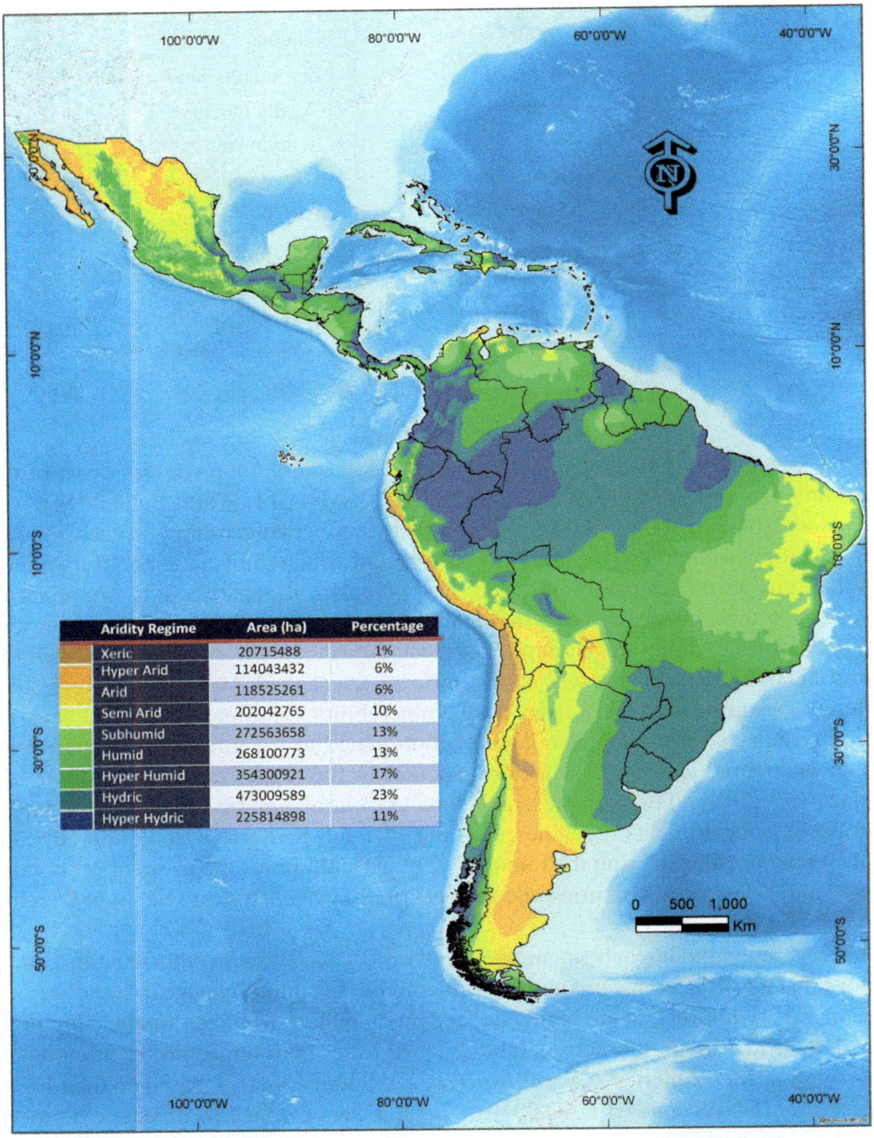

Map 3.1 Aridity map of Latin America and the Caribbean. *Source* UNESCO/PHI-LAC (2010)

3.2 The Climate Change Context

The IPCC Fifth Assessment Report (AR5) concludes that climate change is unequivocally linked to human activities, particularly greenhouse gas (GHG) emissions and land use changes. Changes are observed in all geographical regions. Average air and ocean temperatures are rising and ice and snow cover is decreasing, average sea levels are rising, and weather patterns are changing (IPCC 2013, 2014a).

Projections from the computational climate models used by the IPCC indicate that changes will continue under a range of possible GHG emission scenarios during the twenty-first century. If emissions continue to grow at the current rate, impacts by the end of the century are expected to show an average temperature increase of between 2.6 and 4.8 °C above current temperatures and average sea levels of 0.45–0.82 m above current values (IPCC 2013, 2014a).

To prevent the most severe impacts of climate change, the parties to the UNFCCC agreed on a target of keeping the global average temperature increase to no more than 2 °C above pre-Industrial Revolution levels and considering bringing that target to 1.5 °C in the near future.[3] The contribution to AR5 from IPCC Working Group I (WGI) concluded that by 2011, we had already emitted about two-thirds of the maximum cumulative amount we could emit to have a 66% chance of reaching the 2 °C target. Even if emissions were stopped immediately, temperatures will remain high for centuries due to the effect of atmospheric concentrations of GHGs already in the atmosphere. Therefore, limiting temperature increases will require substantial and sustained reductions in GHG emissions over time (IPCC 2013).

With regard to temperatures, as shown in Map 3.2, only in the RCP[4] 2.6 scenario would the goal of no more than a 2 °C increase in temperature by the end of 2100 be achieved. If a scenario such as RCP 8.5 prevails, as shown, the expected increases in temperatures may exceed 5 °C in some areas covered by this study (IPCC 2013). In the case of changes in rainfall, although some changes are already being observed even in CPR scenario 2.6, there are noticeable reductions in rainfall for the case of RCP scenario 8.5, mainly in the central and southern zone of the Andes, in the NE of Brazil, in Venezuela, Colombia and in much of Central America (IPCC 2013). Both changes in average temperatures and changes in rainfall will have impacts on those economic sectors and ecosystems that depend on climate, as is the case with the agricultural sector (Map 3.3).

---

[3] At the 21st Conference of the Parties (COP21) to the United Nations Framework Convention on Climate Change (UNFCCC), held in Paris (France) in December 2015, the so-called "Paris Accords" were signed in which it was agreed to impose this limit on temperature increase (at the end of the twenty-first century) below 2 °C, compared to the levels observed before the Industrial Revolution.

[4] *Representative Concentrations Pathway* in reference to the atmospheric concentrations reached by that scenario at the end of the twenty-first century.

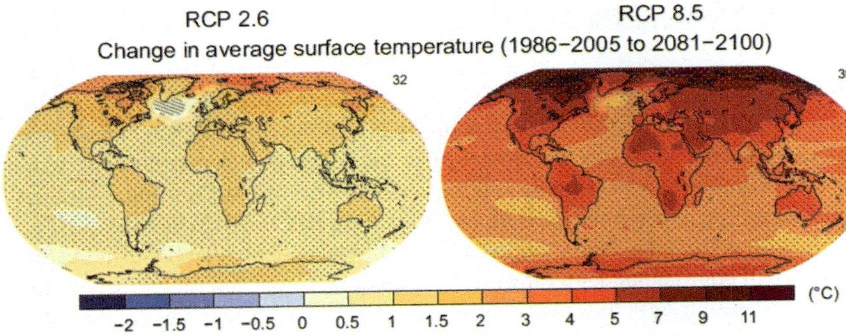

Map 3.2 Changes in average surface temperature toward the end of the twenty-first century for RCP 2.6 and RCP 8.5 scenarios. *Source* IPCC (2013)

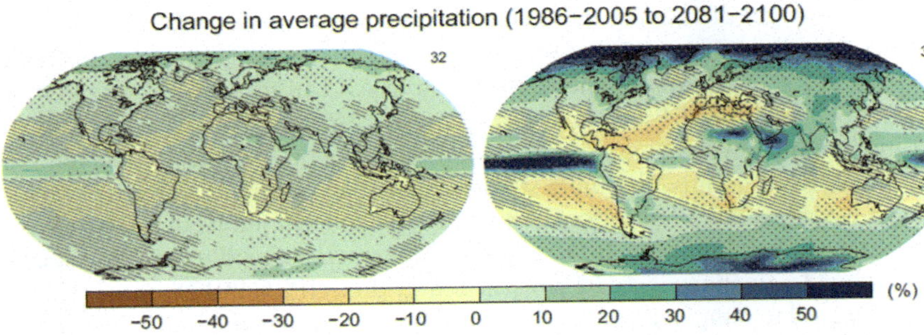

Map 3.3 Changes in average rainfall toward the end of the twenty-first century for RCP 2.6 and RCP 8.5 scenarios. *Source* IPCC (2013)

3.3 Climate Change and Agriculture

The performance of the agricultural sector (and, consequently, food production), has a very strong interrelationship with the prevailing conditions in the climate. Therefore, agricultural activities can be very affected by changes in the climate. In addition, the exposure of these activities to climate change will depend on the situations prevailing in each specific region under analysis. As shown in Table 3.2, several studies have concluded that (a) agricultural activities have great potential to be negatively affected by climate change and (b) exposure to climate change will depend on the region and economic conditions within the countries involved (see also Fig. 3.1).

The latest IPCC assessment report paid particular attention to this issue. In this regard, the main points to be highlighted in the 5AR regarding the potential impact of climate change on agriculture at the global level are the following (IPCC 2013, 2014b, 2015):

Table 3.2 Climate change potential impacts on agriculture

Agriculture and climate change		
Conclusion	Evidence	Author
Agricultural activities have great potential to be negatively affected by climate change	The climate is a determining factor for agricultural productivity and therefore this is one of the sectors most likely to suffer alterations in the face of climate change	Adams et al. (1998) Fischer et al. (2005) Mendelsohn (2009) Tubiello and Rosenzweig (2008) Hertel and Rosch (2010)
	As agricultural income may fall in the face of rising temperatures, rural areas will be most affected	IPCC (2014a, b, c)
	In the face of the potential ravages of climate change, agriculture and the populations that depend on it need to have greater adaptive capacity	McCarl (2010) Mendelsohn (2009)
Exposure to climate change will depend on the region and economic conditions within the countries	Developing countries are more vulnerable to climate change, as they are more dependent on agriculture and experience higher temperatures	Fischer et al. (2005) Mendelsohn (2009)
	Within developing countries (as in Latin America), small farmers within rural areas would be most affected	Birthal (2014) IPCC (2014a, b, c) Mendelsohn (2009) Seo and Mendelsohn (2008)

Source Adaptation from López Feldman (2016)

- Climate change-related impacts are already affecting crop yields in many parts of the world. This trend is expected to continue as long as temperatures continue to rise. Affected crops include staple foods such as wheat, maize and rice.
- Climate change is expected to increase the volatility of agricultural commodity prices and reduce food quality.
- Farmers can adapt to changes to some extent, but there is a limit to the situations that can be managed. Climate impacts are expected to affect poor farmers disproportionately to those who are most resilient. However, adaptive capacity is expected to be exceeded in tropical regions if temperatures rise by about 3 °C or more.
- Ambitious approaches to both adaptation and mitigation of GHG emissions are in the best interests of the agricultural industry.
- GHG emissions from the agricultural sector comprise about 10–12% of total anthropogenic GHG emissions in 2010. The sector is primarily responsible for human emissions of GHGs other than CO_2, such as methane (CH_4) from rice fields

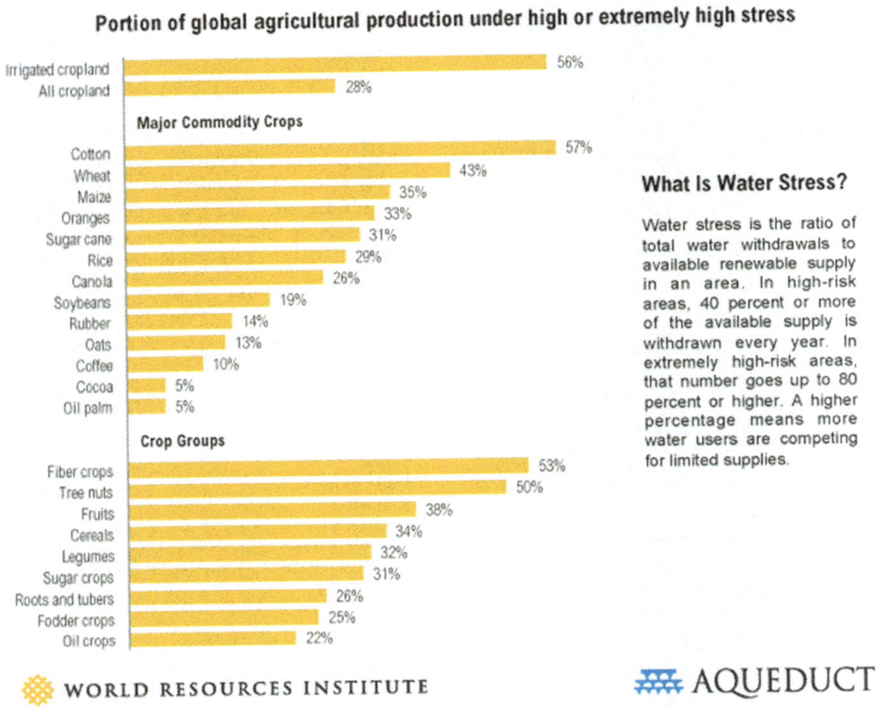

Fig. 3.1 Agricultural production under high or extremely high water stress. *Source* WRI (2016)

and enteric fermentation of ruminants or nitrous oxide (N_2O) from agricultural soils and fertilizer use.

- The agricultural sector offers opportunities for GHG emissions mitigation that include reducing emissions from deforestation processes, soil management, and livestock. Carbon can also be captured and stored in both soils and biomass, and biofuels can play an important role in replacing emissions from fossil fuels.

In this document, however, the analysis will focus on aspects related to the impacts of climate change on agriculture (mainly in certain arid and semi-arid regions of Latin America) rather than on its mitigation opportunities.

The effects of climate change on crops and food production are already evident in several regions of the world, where negative impacts are more common than positive ones. Without adaptation, climate change is projected to reduce production as a result of local temperature increases of 2 °C or more by 2050 (above pre-Industrial Revolution levels), although some specific sites may benefit. After 2050, the risk of more severe effects on crop yields increases and the final outcome will depend on the level of warming. Climate change will be particularly severe on agricultural production in Africa and Asia. Global temperature increases of 4 °C or more, combined with increased demand for food, would pose major risks to the food security of the world and of these regions in particular (IPCC 2013, 2014a, b, c).

The greatest impacts of climate change, in global terms, are expected to be on water availability and supply, food security and income from activities linked to agriculture, including potential displacements from food crop production areas and industrial crops. These are some aspects of a complex picture that involves interactions between CO_2 levels, ozone levels, increases in average, minimum and maximum temperatures, the behavior of extreme temperatures, the reduction of water availability and changes in the nitrogen cycle, whose global effects are difficult to predict. In addition, the areas suitable for coffee, tea, and cocoa cultivation, which support millions of small farmers in more than 60 countries, are likely to be significantly reduced with rising temperatures coupled with other factors. These projected impacts will occur at the same time as demand for beans is expected to increase by 14% per decade through 2050. The risks are greater in tropical countries (IPCC 2014a, b, c, 2015).

According to the most recent studies, the main points to be highlighted are the following (IPCC 2014a, b, c, 2015).

- *Food Security*: Recent extreme weather events (heat waves, droughts, floods, and forest fires) combine with long-term trends (rising temperatures, changes in precipitation patterns), resulting in consequences for the agricultural sector and global food security. Different terrestrial ecosystems that provide many vital services for agricultural production (including nutrient cycles, waste decomposition, and seed dispersal) will be damaged and may even be lost due to climate change. Climate change is also the second most important global threat to insect pollination, after habitat loss, with consequent consequences for agriculture. See Fig. 3.1.
- *Crop Yields*: Climate change is already affecting wheat, maize, and rice yields, in most regions and differently in different cases. Without adaptation, a temperature increase of 2 °C or more could further reduce yields. While an abundance of CO_2 atmospheric, in most cases, has a stimulating effect on photosynthesis and consequently on plant growth, the presence of greater amounts of tropospheric ozone reduces growth. High ozone levels are likely to cause losses of about 10% in crops such as wheat and soybeans.
- *Developing Countries*: The risks to agriculture from changes in climate are particularly acute in developing countries. The vulnerability of farmers and pastoralists exposes them to a lack of fundamental resources for resilience such as access to various resources (financial, technological, knowledge). In addition, climate-related risks interact with existing environmental stressors such as biodiversity loss, soil erosion, and water pollution, and with social stressors such as inequality, poverty, gender discrimination, and lack of institutional capacity. These interactions constitute additional risks to agricultural production and food security.
- *Water Supply Security and Availability*: In many regions, hydrological systems are being altered as a result of changes in precipitation levels and patterns, snow and ice melt, and glacier retreat. All of this will affect the availability and quality of water resources. Climate change could significantly reduce the availability of

surface water and groundwater resources, mainly in the more arid subtropical regions. Each degree of increase in temperature is expected to reduce renewable water resources by at least 20% if it occurs at the same time as a 7% growth in the world's population.

- *Agricultural Price Volatility*: An important factor in recent food price increases was the increase in demand for certain crops driven, in many cases, by increased land allocation for biofuel production. However, the impact of climate on food production also played a major role in price fluctuations, with recent price spikes due to extreme weather events in major producing countries. Prices of rice (37%), maize (55%), and wheat (11%) are projected to rise by 2050 due to the additional stress of climate change. Greater price volatility generates uncertainty, and rising food prices (in this case, climate-related) have a disproportionate impact on the well-being of the poor, increasing their vulnerability. It is estimated that the 2010/2011 food price increase will have pushed 44 million people below the basic needs poverty line in 28 countries (IPCC 2014a, b, c).
- *Food Quality*: The quality of some foods is likely to be affected. Growing wheat, rice, barley, or potatoes in high CO_2 concentrations reduces the protein content by 10–14%. Some crops may also show reductions in mineral and nutrient concentrations.
- *Pests and Diseases*: Some pest outbreaks are attributed to climate change. The increase in the earth's temperature, changes in precipitation patterns, and the increase in the frequency and intensity of extreme heat undermine the natural regulation of pests and diseases. This in turn can lead to the loss of important environmental services and facilitate increased dominance of harmful invasive organisms. Projected increases in pest damage to crops are expected to further affect food production and raise the cost of key raw materials.
- *Livestock*: Increased heat stress coupled with more frequent extreme weather events will have negative consequences for livestock. Improved high-yielding varieties are particularly at risk. In developing countries, landraces tend to be more tolerant to heat and seasonal malnutrition. Hazardous livestock pathogens are expected to expand their geographical range as a result of climate change.
- *Employment*: Labor productivity in the agricultural sector, particularly for manual work in humid climates, is likely to fall as a result of heat stress and vector-borne diseases.
- *Supply Chain*: Food production is only one part of the agricultural supply chain. The sector also depends on refrigeration, transport, processing, and retailing. Each of these links in the chain is exposed to climate risks, such as disruption of operations and the need for greater temperature control.

However, farmers and other actors in the food production chain also have the potential to adapt to some of the impacts of climate change (IPCC 2014b, 2015; University of Cambridge 2014). Nevertheless, it is difficult to generalize about the actions that are available because adaptation is highly "context-specific", so there is no simple approach to risk reduction that can be applied across all regions, sectors, and contexts. One consensus is that the capacity of the agricultural sector to cope with climate events is likely to decline over time if the climate becomes warmer, beyond certain temperatures. Strategies for effective, resilient, and sustainable grain production include improving knowledge of crop seasons, improving crop rotation systems, adaptive water management techniques, and improving weather forecasting (IPCC 2014b, 2015; University of Cambridge 2014).

3.4 Climate Change and Agriculture in Latin America

As stated above, the impact of climate change on the various regions, socioeconomic sectors, and ecosystems will depend on the specific characteristics of these elements and the context in which these changes occur.

In this sense, Map 3.4 shows a summary of the main changes observed, both in climate and in other environmental factors, in the regions of Central and South America, according to the information gathered and presented in the 5AR of the IPCC WGII. The most relevant issues that deserve to be highlighted, referring to the observed trends in the region's climate and the potential impacts of climate change on agricultural activities, natural ecosystems, potential land uses, and the vulnerability of the populations involved, in the scope of Central and South America, are the following:

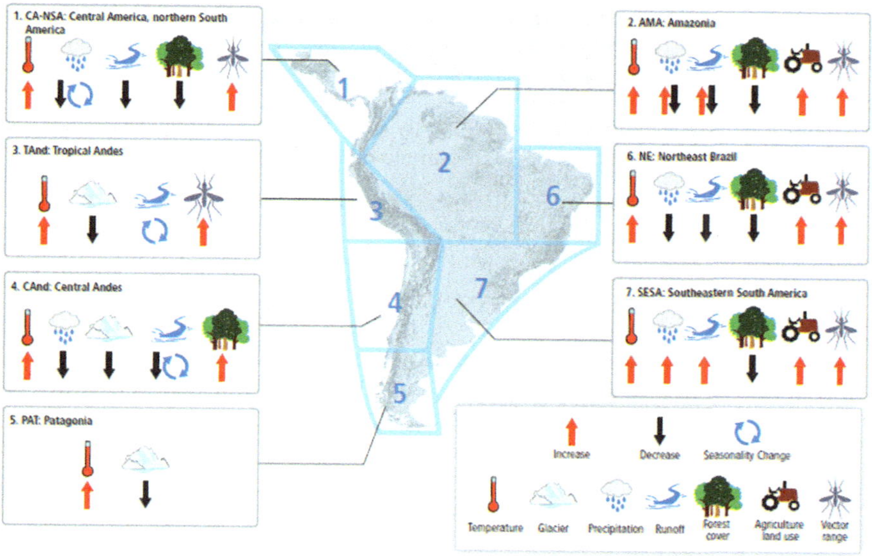

Map 3.4 Summary of the main changes observed in climate and other environmental factors in Central and South America. *Source* IPCC (2014a, b, c)

- Significant trends in both temperature and rainfall in Central and South America (high confidence level).[5]
  - Increase in annual rainfall (SESA),[6] mainly during the period 1950–2008 Downward trends in Central America and south-central Chile (and on the other side of the Andes).
  - Temperature increase (since mid-1970s) in Central and South America, except for a cooling on the Chilean coast.
- Climate variability and extreme events have severely affected the region (medium confidence level).
  - Increases in temperature extremes (Central America and most tropical and subtropical regions of South America).
  - The more frequent extreme rains in the SESA have favored landslides and flash floods.
- Changes in flows and water availability. This is projected to continue in the future (Central and South America) and will affect regions that are already vulnerable (high confidence level).
  - Andean glaciers are retreating (affecting the seasonal distribution of flows).
  - Increases in runoff in the rivers of the Plata Basin and decreases in the central Andes (Chile and Argentina) and in Central America in the second half of the twentieth century, which are related to changes in precipitation.
  - The risk of water supply shortages will increase due to reductions in precipitation and increases in evapotranspiration in semi-arid regions. This will affect urban water supply, hydropower generation, and agriculture.
- Land use change contributes significantly to environmental degradation, which exacerbates the negative impacts of climate change (high confidence level).
  - Deforestation and land degradation are mainly attributed to increasing agriculture (both intensive and extensive).
  - Agricultural expansion (which in some regions is associated with increases in rainfall) has affected fragile ecosystems such as the edges of the Amazon forest and the tropical Andes.
  - Deforestation rates in the Amazon have decreased substantially since 2004 but not so in the *Cerrado* which still has high levels of deforestation (average rate of 14,200 km^2 per year for the period 2002–2008).

---

[5] The IPCC assigns a degree of certainty to each major conclusion based on the type, quantity, quality, and consistency of the evidence and the level of consensus among scientists. The terms used to describe the evidence are: limited, medium, or strong. To describe the level of agreement: low, medium, or high. In 5AR, when we talk about the "level of confidence" in a conclusion, it is derived from the synthesis between the existing evidence and the degree of scientific consensus regarding what that evidence means. The confidence levels assigned by the IPCC are: very low, low, medium, high, and very high (see IPCC 2013, 2014b).

[6] South East of South America.

- Conversion of natural ecosystems is the major cause of biodiversity and ecosystem loss in the region and is a major cause of anthropogenic climate change (high confidence level); climate change is expected to increase the rate of species extinction (medium confidence level).
  - Brazil: Movement of birds and plants toward the south.
- Economic conditions improved in the region between 2007 and 2014. However, in most countries there is still a high and persistent level of poverty resulting in high vulnerability and increased risk to climate variability and change (high confidence level).
  - Poverty levels remain high in all countries.
  - HDI varies from Chile and Argentina to Nicaragua and Guatemala.
  - Economic inequality translates into inequality in access to water, sanitation and adequate housing (especially for the most vulnerable groups), which leads to a low capacity to adapt to climate change.
- Projections to 2100 in Central and South America: Temperature increases along with precipitation increases in some areas, and precipitation decreases in other areas of the region (medium confidence level).
  - For RCP 4.5 and 8.5, the heating in Central America is between + 1.6 and + 4.0 °C and between + 1.7 and + 6.7 °C in South America both by 2100 (medium confidence level).
  - Changes in rainfall are between − 22 and + 7% by 2100 in Central America. In South America, precipitation varies geographically (increases in the Southeast and decreases in the West and Northeast): − 22% in the Northeast of Brazil and + 25% in the SE of the continent (low confidence level).
  - Increases in drought periods (tropical South America east of the Andes) and in the number of warm days and nights in most of South America (medium confidence level).
- Sea level rise and human activities in marine and coastal ecosystems pose threats to fish stocks, corals, mangroves, recreation, tourism, and disease control (high confidence level).
  - Between 1950 and 2008 MSLR[7] varied between 2 and 7 mm per year.
  - The main drivers of mangrove loss are deforestation (and consequent conversion of land to agricultural use) and shrimp farming ponds.
- Wide spatial variability of changes in agricultural productivity associated with climate change, with implications for food security (medium confidence level).
  - In SESA, projections indicate more rainfall and average productivity could be maintained or even increased until the middle of the century (average confidence level).

[7] Mean Sea Level Rise.

  - In Central America, Northeast Brazil and parts of the Andean region, temperature increases and rainfall decreases could reduce productivity in the short term (2030), threatening the food security of the poorer population (medium confidence level).
  - South America will be a key region for food production in the future. Challenge: to increase the quality and production of food and bioenergy, while maintaining environmental sustainability under conditions of climate change. Adaptation measures include crop, irrigation, and water use management along with genetic improvements (high confidence level).

- Bioenergy production has a potential impact on land use change (mainly deforestation) and could be affected by climate change (level) medium confidence.
  - Sugarcane and soybean are likely to respond positively to increases in CO_2 and temperature concentrations with increases in production and yields, even in the face of a decrease in water availability (high confidence level).
  - The expansion of sugarcane, soybean, and African palm crops may have effects on land use, leading to deforestation of parts of the Amazon and Central America and, in some countries, could lead to job losses (medium confidence level).

From the point of view of the specific impacts on agricultural activities, food security, and the vulnerability of the region's populations and ecosystems to climate change, the combination of stress factors originating in both climate and human issues operate on various elements and in various forms, depending on the countries, regions, ecosystems, economic sectors, and actors involved. Some important aspects to analyze, in the case of South and Central America, are the following

- *Expansion of the area planted for food and biofuel production*: In both South and Central America, there is a significant increase in agricultural production (associated with the expansion of planted areas), resulting from the increase in global demand for food and biofuels. This trend is expected to continue in the future. Ecosystems are affected by climate variability and change, along with changes in land use, which are driving environmental changes. In both regions, even under conditions of climate change, optimal land management could efficiently combine agricultural and biofuel production with ecosystem preservation. However, current practices are leading to a deterioration of ecosystems throughout the continent. In the case of ES, it can lose between 1 and 21% of its arable land as a result of climate change and population growth (IPCC 2014c).
- *Changes in crop productivity*: In SESA there were significant increases in rainfall and wetter soil conditions during the twentieth century, benefiting productivity (summer crops and pasture), contributing to the expansion of agricultural areas. In Argentina, SE Brazil and Uruguay, maize and soybean yields increased (9–58%) due to the more humid conditions observed between 1970 and 2000 compared to 1930–1970. SESA's agricultural systems could be threatened if drier conditions occur, as a result of interdecadal variability, even if rainfall estimates increase

by 25% by 2100. This could jeopardize the viability of continuing to farm in the marginal areas of the Pampa and Chaco in Argentina. Warming (since 1981) has reduced the productivity of corn, wheat, and barley (globally), but the drops in yields were small compared to the technological gains in yields over the same period. In central Argentina, wheat yields have been declining more and more. Between 1980 and 2008, changes observed in temperatures and rainfall during the growing season slowed the trend toward higher yields due to genetic improvements in Brazil (soybeans, maize and wheat), as well as in Paraguay (soybeans). In contrast, rice in Brazil and soybean in Argentina have benefited from the trends of higher rainfall and temperatures. In Argentina, increases in soybean yields (see Argentina 2007, 2015; ECLAC 2014a) may be associated with types of climate that favor the entry of cold air from the south, reducing heat stress during the flowering and pod growth period, and climates that increase the likelihood of dry days in the harvest (IPCC 2014c).

- *Food production, agricultural commodities, and food security*: The effect on food production and food security of future climate scenarios shows a wide range of uncertainties. One of them is the effect of increased CO_2 on plant physiology. Many crops (soybeans, beans, maize, and sugarcane) are likely to respond with increased productivity. However, food quality may fall due to higher sugar content in grains and fruits and drops in protein content in cereals and legumes. Uncertainties (associated with climate and crop models, but also with human behavior) make long-term predictions of food production difficult. It is expected that changes in food production (and consequently their impact on food security) may have great spatial variability. Average productivity in SESA could be maintained or even increased until the middle of the century, although year-to-year and decadal climate variability is likely to cause significant damage. In other regions, such as Northeast Brazil, Central America, and some Andean countries, agricultural productivity could fall in the short term, threatening the food security of the poorer population. The expansion of pastures and farmland is expected to continue in the coming years, especially from a growing global demand for food and biofuels. The great challenge for Central and South America will be to increase the production of both food and bioenergy while maintaining environmental quality, in a scenario of climate change.
- *Impacts on agriculture, disaggregated by areas*: Differences are presented according to the areas concerned:

  - In Uruguay and Argentina, productivity may increase or remain stable until the middle of the twenty-first century, although climate variability from one year to the next and from one decade to the next may cause significant damage. Warmer and more humid conditions can benefit crops in the east and south of the Pampas plains (see ECLAC 2014a). In Paraguay, soybean, corn, and wheat yields will vary greatly until 2020 (ECLAC 2014b). In Chile and western Argentina, yields may fall due to limitations on water availability.
  - In southern Brazil, yields of irrigated rice are expected to increase. If technological improvements are considered, bean and corn productivity could

increase from 40 to 90%. Sugarcane production could benefit if warming allows the expansion of planted areas to the south, where low temperatures are a constraint. In the State of São Paulo, until the middle of this century, increases in crop productivity could rise by up to 6%. In Northeast Brazil, declines in yields are projected for subsistence crops such as beans, corn, and cassava. In addition, increases in temperature may reduce the areas currently favorable to cowpeas. The greater warming forecast for the end of the century could make coffee crops unviable in São Paulo and Minas Gerais if these higher temperatures are not accompanied by adaptation measures. Coffee crops will have to be moved to southern Brazil, northern Argentina and Brazil's border with Uruguay. Brazilian potato production may be restricted to a few months in the currently warm areas that still allow annual potato production.
- In Central America, Northeast Brazil and the Andean region, climate change may affect crop yields, local economies, and food security. By the end of the century, temperatures during the growing season in the tropical areas of South America, east of the Andes, and in Central America are likely to exceed the seasonal extremes documented between 1990 and 2006, affecting the productivity of regional agriculture and human well-being. Also, in Central America, the projected changes in climate can severely affect the poorest populations, mainly in their food security, increasing the current rate of chronic malnutrition. This is especially serious in Guatemala, which is currently the country with the lowest food security as a percentage of the population (30.4%) and where the problem has been growing in recent years. The impact of climate variability and change is a major challenge for the region. As an example, the recent rust problem in the coffee sector (2012–2013) affected nearly 600,000 ha (55% of the total area) and reduced employment by 30–40% in the 2013–2014 harvest. At least 1.4 million people in Guatemala, El Salvador, Honduras, and Nicaragua depend on the coffee sector which is very susceptible to climate variations. In Panama, high year-on-year climate variability will continue to be the dominant influence on seasonal corn yields in the coming decades. In the future, warmer conditions combined with greater variability in rainfall may reduce productivity of maize, rice, and beans.[8] Wheat and rice yields may fall by up to 10% by 2030.

- *Impacts of climate change on food prices*: The impacts of climate change on the region's well-being can be felt not only in crop yields but also in the price of food and export goods. While a price increase may benefit those countries that are net exporters (Argentina, Brazil, Paraguay, Uruguay), other countries may be affected. Many poor households can be badly affected by food price increases. An increase in agricultural prices in the period 2007–2009 was the main explanation for the increase in poverty in Nicaragua (IPCC 2014c).
- *Extreme weather events*: Some IPCC reports (AR4 and SREX, for example) contain extensive evidence of increased extreme weather events in Central and

[8] In Central America, nearly 90% of agricultural production for domestic consumption consists of corn (70%), beans (25%), and rice (6%) (IPCC 2014c; ECLAC 2010).

South America. During the period 2000–2013, there were 613 extreme weather and climate events that led to 13,883 deaths and 53.8 million people affected, with losses estimated at US $52.3 billion. Between 2000 and 2009, 39 hurricanes occurred in the Caribbean Basin and Central America, compared to 15 and 9 in the 1980s and 1990s, respectively. In SESA, more frequent and intense extreme rainfall has led to an increase in the occurrence of flash floods and landslides. In the Amazon, extreme droughts were reported in 2005 and 2010, while unprecedented floods were observed in 2009 and 2012, and in 2012–2013 an extreme drought affected NE Brazil.

- *Ecosystems*: Central and South America are home to unique ecosystems and a wealth of biodiversity. However, natural ecosystems are affected by climate variability and/or change as well as by land use change. Deforestation, land degradation, and biodiversity loss are mainly caused by an expansion of the agricultural frontier to increase agricultural and livestock production for traditional export activities and bioenergy crops. Agricultural expansion has affected fragile ecosystems, causing serious environmental degradation and the reduction of environmental services provided by these ecosystems. Deforestation has intensified the process of soil degradation, increasing the vulnerability of communities exposed to floods, landslides, and droughts. Plant and animal species (especially amphibians) are declining rapidly in Central and South America. However, the region still has large areas of natural vegetation cover, with the Amazon as the main example.

In the following sections of this document, the analysis will concentrate on four areas considered key from the point of view of agricultural production in both Central and South America: The Gran Chaco Region, the Brazilian Northeast, the Venezuelan Coast and the Central American Semi-Arid Belt.

3.5 The Gran Chaco Region (Argentina-Bolivia-Paraguay)

The American Gran Chaco Region is an environmental unit that extends from definitely tropical latitudes (18° S) to clearly subtropical environments (31° S), ranging from 57 to 66° West (see Map 3.5). It constitutes a great plain of approximately 1,140,000 km^2 distributed in the Center-North of Argentina (including the totality of the provinces of Formosa, Chaco, Santiago del Estero and part of the provinces of Salta, Jujuy, Tucumán, Santa Fe, Corrientes, Córdoba, Catamarca, La Rioja, San Luis, and San Juan), West of Paraguay (mainly the departments of Alto Paraguay, Boquerón and *Presidente Hayes*), SE of Bolivia (occupying mainly part of Tarija, Chuquisaca, and Santa Cruz), and a small part of South Brazil. The area of the Chaco corresponding to Argentina is 673,000 km^2, to Paraguay 269,000 km^2, to Bolivia 124,000 km^2 and the remaining 8000 km^2 are located in Brazil. See Map 3.6 (REGATTA 2014).

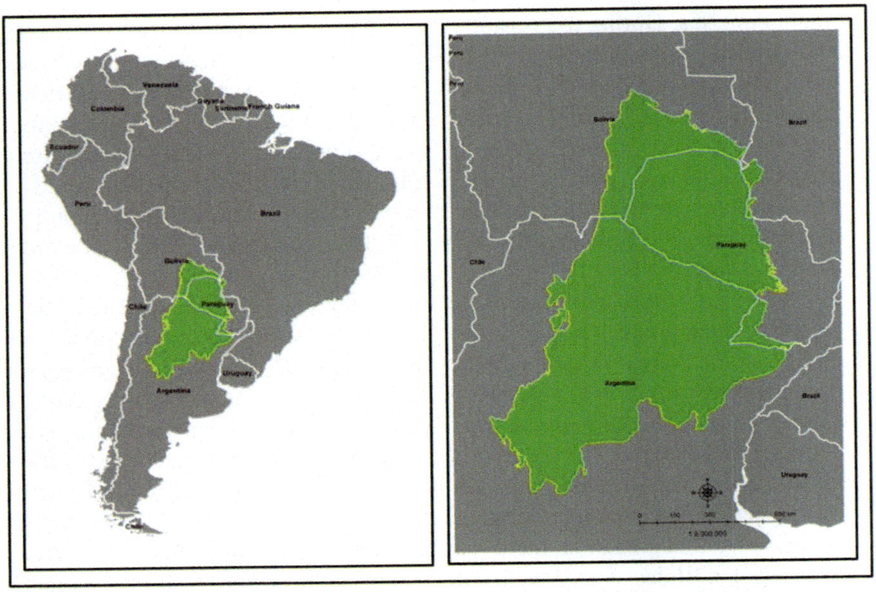

Map 3.5 Location of the American Gran Chaco Region in the context of South America and the region. *Source* REGATTA (2014)

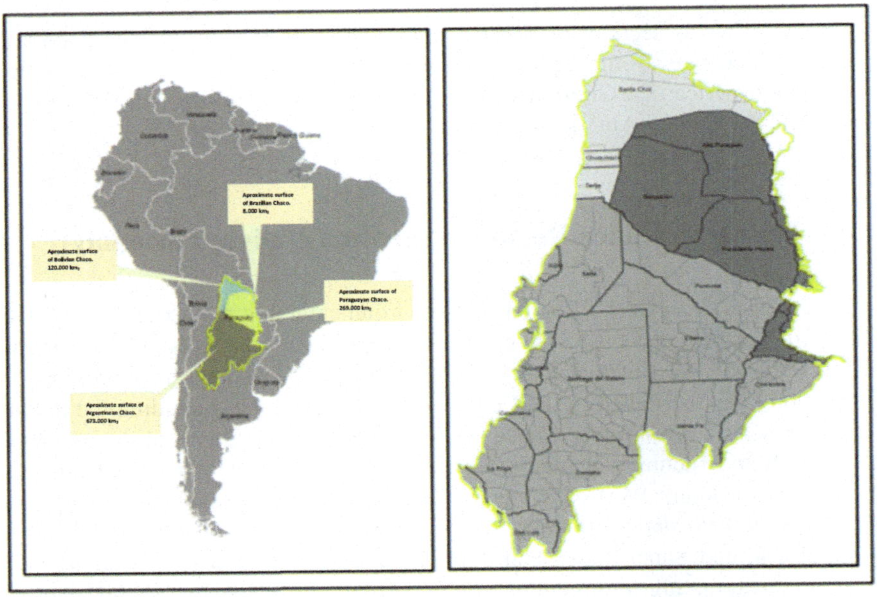

Map 3.6 Area corresponding to each country part of the American Gran Chaco Region and corresponding administrative units. *Source* Adapted from REGATTA (2014)

Among the various criteria that exist to classify this ecosystem, the most accepted is that which defines it in terms of three subzones: Subhumid Chaco, with rainfall ranging from 1200 to 700 mm; Semi-Arid Chaco, with 700 to 500 mm; and Arid Chaco, with rainfall from 500 to 300 mm. Rainfall has a marked summer concentration, and consequently, dry winters occur, with records of the driest quarter in the order of 10–12% in the Humid Chaco, while in the Dry Chaco and Arid Chaco, they only reach 1–5%. Thus, the rainfall regime is the main factor in defining the regions and productive activities. In addition, the region presents marked climatic gradients, with average annual temperatures ranging from 18 to 26 °C, while potential evapotranspiration varies from 900 mm in the South to 1600 mm on the border between Paraguay and Bolivia (REGATTA 2014).

The total annual rainfall in the Gran Chaco (Map 3.7), shows two minimums, from a spatial point of view: one located on the border between Bolivia and Paraguay (with an annual minimum of 500 mm), and another over the province of La Rioja, Argentina (of 400 mm). Precipitation increases in an easterly direction with maximum values of 1200 mm per year (REGATTA 2014).

The seasonality of rainfall is also very variable. There are clear differences between the Humid Chaco and the Semi-Arid Chaco. There is a well-defined rainy season. Summer is the season of highest rainfall with minimum accumulated values of 200 mm in the *Boqueron* department in Paraguay and in La Rioja, Argentina, while the highest values (around 400 mm) are found on the eastern and western

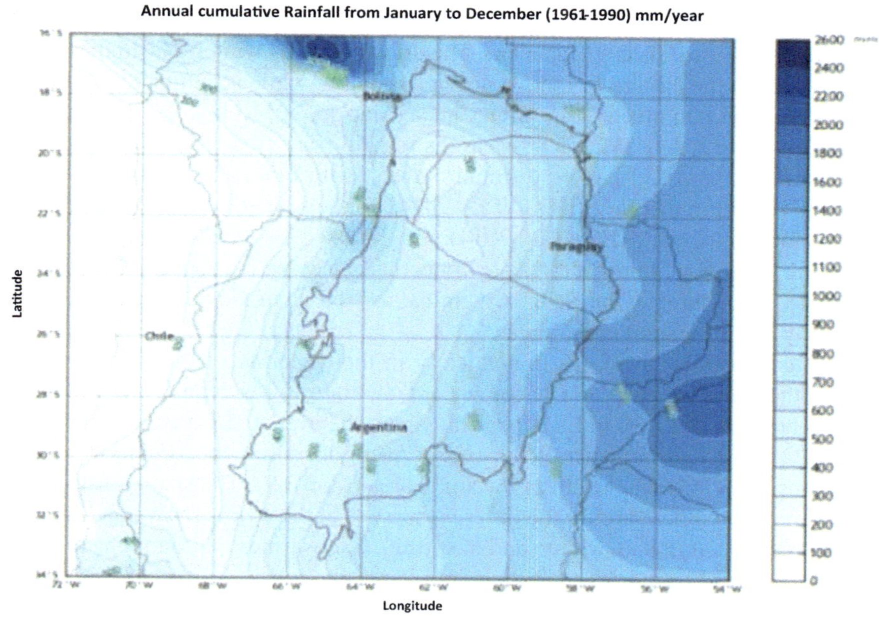

Map 3.7 Annual rainfall in the American Gran Chaco. *Source* Adapted from REGATTA (2014)

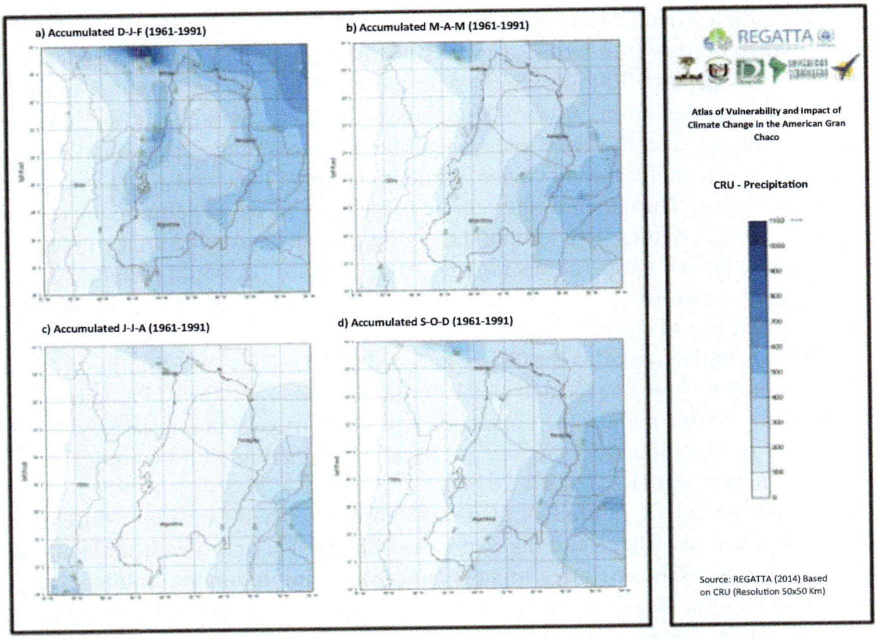

Map 3.8 Seasonal distribution of rainfall. *Source* Adapted from REGATTA (2014)

borders of the Gran Chaco (Map 3.8a). The driest season is winter. The highest value is recorded on the eastern border, at around 100 mm, while accumulated rainfall is less than 100 mm in much of the Gran Chaco (Map 3.8c). Meanwhile, in the autumn (Map 3.8b) and spring (Map 3.8d) seasons rainfall transitions from the high to the low rainfall season and has very similar values, with ranges from 300 mm in the East to 100 mm in the West (REGATTA 2014). In the region, water is not a limiting factor, nor has it been (REGATTA 2014).

As far as average temperatures are concerned, in some places they reach values above 28 °C in summer (in Salta, Argentina, as well as in the departments of Alto Paraguay and Boquerón in Paraguay).

The temperature decreases dramatically in the direction of the Cordillera and toward the south of the Gran Chaco, where temperatures reach an average of 21 °C (with a hot core of 27 °C in La Rioja). In winter, the average temperature is slightly above 22 °C in the northeast of the Gran Chaco, decreasing according to latitude decreases to 13 °C in the south of the study area (Map 3.9). The spatial distribution of the average temperature for the different seasons of the year is shown on Map 3.10 (a. summer, b. autumn, c. winter, and d. spring). The warmest region in the different seasons is in the territories of Bolivia and Paraguay, where it exceeds 28 °C. The coldest area in the Gran Chaco is in the province of Córdoba.

The Chaco region is of great importance from a biological point of view. Not only is it an important center for the dispersion of species (carob trees, *quebrachos*),

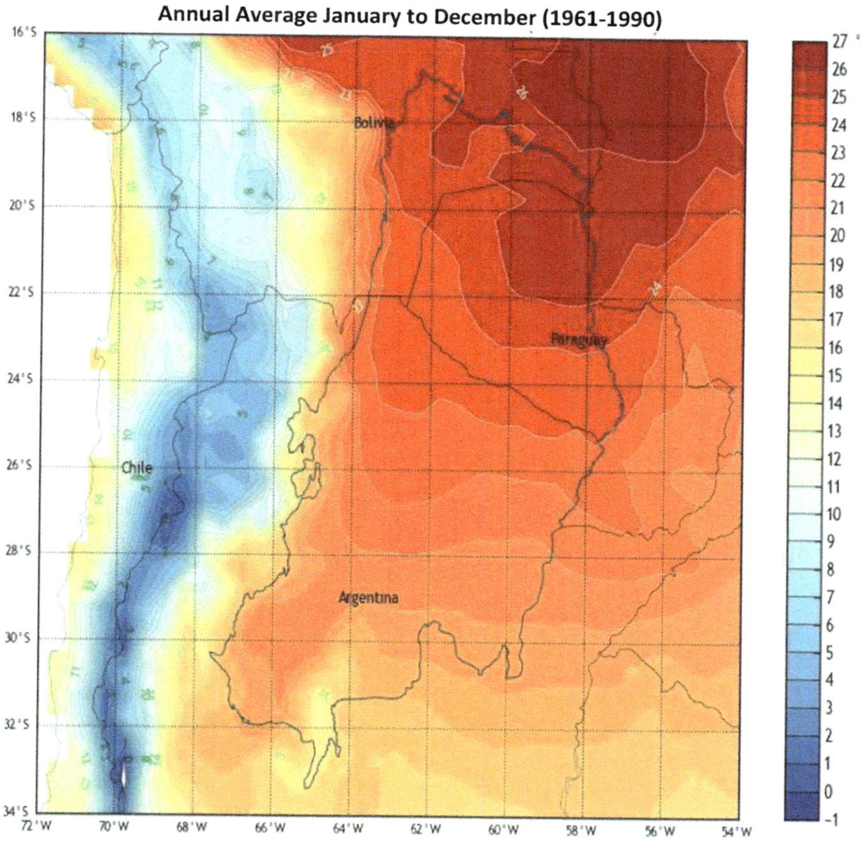

Map 3.9 Annual average temperatures. *Source* Adapted from REGATTA (2014)

but it also has a high level of representation of the vegetation of the countries that make it up and contains a significant number of endemic forms. In addition, beyond its biophysical characteristics, the Chaco region has a great diversity and social and cultural complexity. In addition to a very rich mosaic of indigenous cultures (around twenty-six ethnicities), a strong migratory process with communities of diverse origin has historically converged in the area (REGATTA 2014).

The Gran Chaco, like any environmental unit, has natural boundaries that do not coincide with the political boundaries established by nations, as the natural components of this region are shared across political boundaries. Nevertheless, knowledge of the political organization is fundamental when carrying out the methodological proposal and the analysis of the results, as well as when considering the countries' policies and the actions to be recommended at regional and national level (REGATTA 2014).

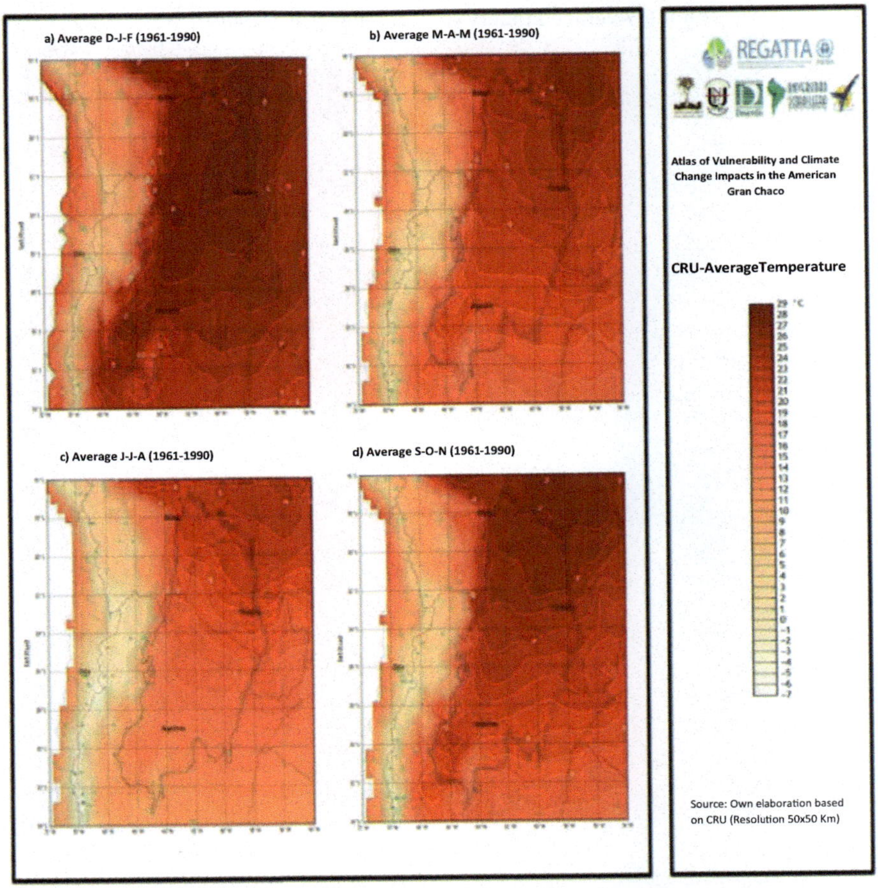

Map 3.10 Average temperatures according to season. *Source* Adapted from REGATTA (2014)

Human presence in this region dates back approximately 7000 years. The original peoples have developed a culture closely associated with their natural resources, being home to nomadic groups of hunter-gatherers, fishermen, and some groups of sedentary farmers. Today there is a (multicultural and multiethnic) population of approximately 8,900,000 people. The highest population concentrations are found in the peripheries of the American Gran Chaco, in territories corresponding to Bolivia and Argentina (where the density reaches more than 40 inhabitants per km^2), there are other areas where the average population density ranges from 10 to 22 inhabitants per km^2; up to the Paraguayan Chaco, where the population density does not exceed the inhabitant per km^2 (REGATTA 2014). The highest concentrations are in urban centers. The distribution of the rural population is closely related to the modes and forms of production, mainly agriculture. In Paraguay and Bolivia there is still a high percentage of rural population (63% and 53% respectively), while in Argentina the technification of the countryside on the one hand and the lack of services and infrastructure in rural areas on the other have caused internal migration processes, leading rural workers to seek employment and better living conditions in urban centers (REGATTA 2014).

The Gran Chaco Americano is mostly comprised of the Rio de la Plata Basin, with a portion to the North belonging to the Amazon River, Rio Grande and the Izozog marshlands (Bolivia), and another portion to the South corresponding to the endorheic basin of the *Salinas Grandes* and the Mar Chiquita Lagoon (Argentina) (REGATTA 2014).

The impacts on the region can be differentiated between those generated by changes in land use over large areas (for pasture, livestock, agriculture, and urban areas) and those caused by extreme weather (droughts and floods). The Gran Chaco has long been an inaccessible, rugged biome with 69.8% natural coverage, even today. However, in recent years its rate of transformation has been increasing, reaching peaks of deforestation of 1500–1800 ha deforested per day. In the Paraguayan Chaco alone, between 200,000 and 280,000 ha of forest have been lost per year in recent years (2008–2011) (REGATTA 2014). In the case of the Argentinean Chaco, the annual deforestation rate between 1972 and 2007 has been around 56,500 ha per year (ECLAC 2014a).

The sites where deforestation took place between 1 and 31 August 2012 can be seen on Map 3.11, marked in red, giving an indication of the dynamics of this process. Map 3.12 shows the land cover in the region in 2012, with green areas corresponding to vegetation, red areas corresponding to agricultural and livestock enterprises or urban use, and yellow areas corresponding to savannahs and wetlands.

In the case of Bolivia, the expansion of the agricultural frontier is mostly associated with industrial crops (soya, rice, wheat, corn, and sugar cane), associated with the conversion of forest areas to agriculture and strongly concentrated in the Department of Santa Cruz. Soil use pressure and inadequate soil conservation practices are deteriorating large areas of agricultural land. If this trend continues, it will reduce the already scarce areas of agricultural suitability in the country, estimated at 2.6% of the total (Bolivia 2001). In addition, the Bolivian Chaco includes areas which according to the Holdridge Life Zone classification would be classified as tropical dry forest and subtropical dry forest. These same areas would become tropical dry forest and very dry tropical forest by 2030, as a result of a drop-in rainfall and an increase in temperature, and this trend would continue until 2100 (Bolivia 2001). In subsequent studies related to the Bolivian Chaco, a reduction in the number of rainy days, an increase in the periods without rain in the cultivation season, greater frequency and intensity of droughts and falls in river flows are predicted. This in turn would result in greater competition for water use, losses in biodiversity, more frequent heat wave events during the summer, a deepening of soil erosion and desertification and a higher degree of contamination of water sources (Bolivia 2009).

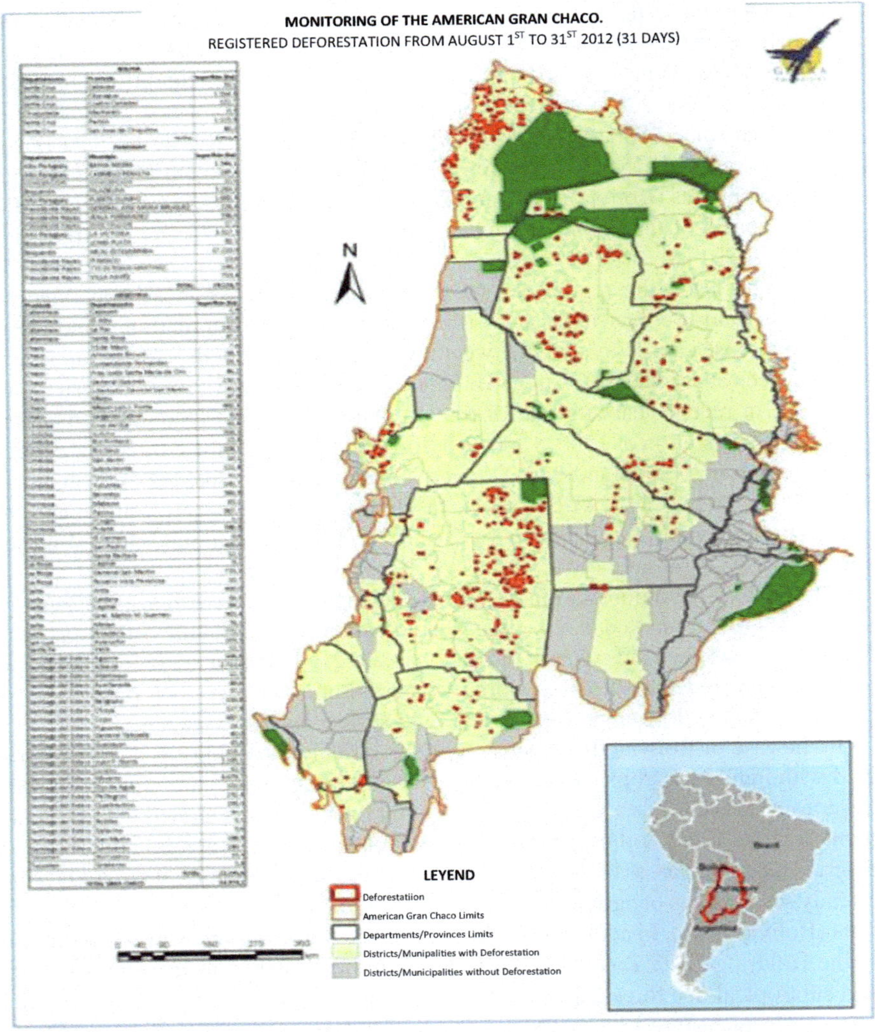

Map 3.11 Deforestation in the American Gran Chaco for a period of one month between 1 and 31 August 2012. *Source* Adapted from REGATTA (2014). Based on Guyra Paraguay

The floodable areas of the South American Chaco correspond to the flood zones of the Pilcomayo, Negro, Paraná, and Paraguay rivers (in addition to other minor watercourses). Map 3.13 allows this information to be crossed with the political subdivisions so that the vulnerability of the different territories (ecological or political) to flooding is more clearly explained.

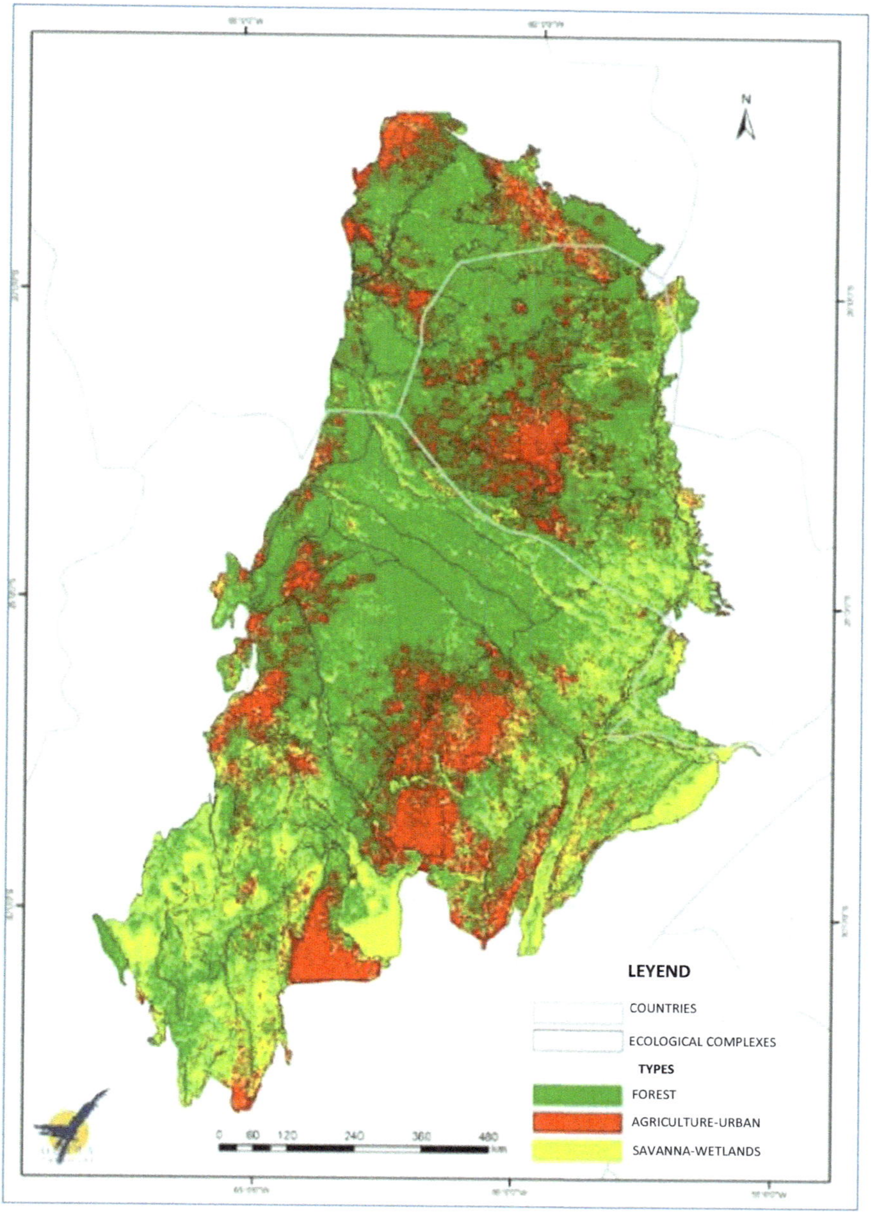

Map 3.12 Soil coverage as of 2012. *Source* Adapted from REGATTA (2014). Based on Guyra Paraguay

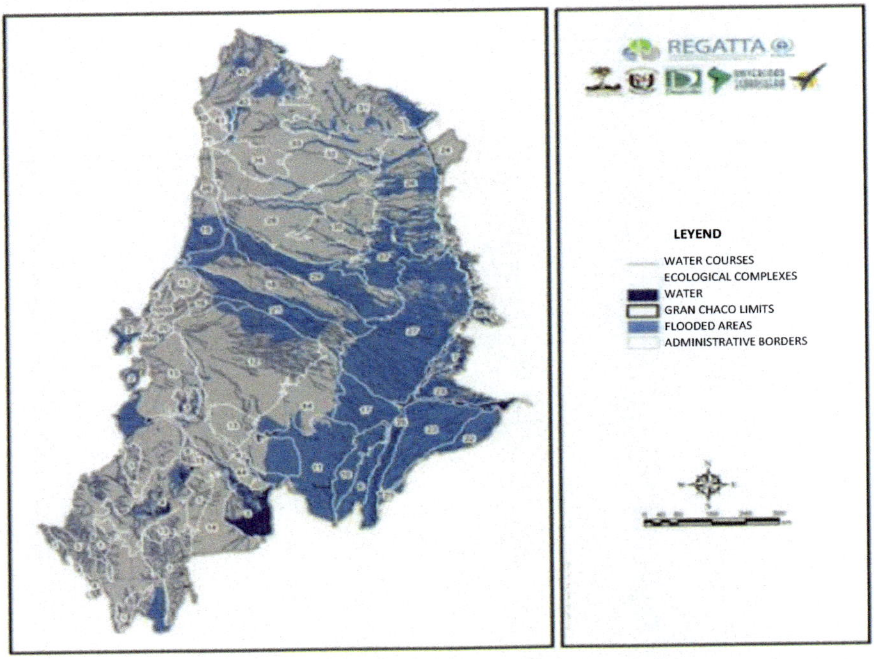

Map 3.13 Floodable areas. *Source* Adapted from REGATTA (2014)

With regard to food production, the region is considered key at the global level and highly competitive due to the incorporation of production and management technologies, in a context where global food demand has undergone major changes in recent decades, mainly influenced by the growth of the Chinese and Indian economies. This increased demand for energy and protein-rich food generates a scenario of growing opportunities for agricultural and livestock production in the region, which is characterized by a wide variety of products (soybeans, sunflowers, corn, wheat, meat, peanuts, among other products) due to climate and ecological differentiation, different patterns of space occupation, and development of different production systems by different population groups (REGATTA 2014; ECLAC 2014a, b; Bolivia 2009).

Two agricultural systems can be distinguished: On the one hand, the traditional or family system, characterized by the small size of the farms, mixed production for consumption and sale, and the use of traditional practices. This form of production currently faces serious problems, among whose causes are the low incorporation of good soil management practices, the low level of technology, as well as its greater vulnerability to climate variability. On the other hand, there is an intensified and less diversified system. This system is linked to markets with greater added value and to the incorporation of new technologies and a more intensive use of information. Crops in the region can clearly be divided into income categories (REGATTA 2014) and self-consumption categories. The income crops, such as cotton, soybean, wheat, rice or sugar cane, have undergone an important technification process in recent years (including the application of direct sowing practices, the use of genetically modified materials and a high level of mechanization, among others). In addition, the cultivated area of the income items has been increasing, advancing on new areas and, in some cases, on pastures. For example, soybean cultivation in Paraguay has increased from 1,050,000 to 3,000,000 ha since 1997, including the presence of experimental soybean crops in the Paraguayan Chaco, where this crop had never been cultivated before (REGATTA 2014).

In the livestock sector, most farms continue to maintain an extensive system of livestock production, mainly cattle, although some small producers are incorporating agroforestry practices that allow them to reduce the effects of the high temperatures characteristic of the region (REGATTA 2014).

Climate change could strongly affect agricultural production in this region, where impacts such as increased water requirements due to rising temperatures or the migration of agroecosystems from their original areas to new agricultural zones are expected. Climate variability events are responsible for significant losses in agricultural production. In this sense, one of the factors that influence food availability is the high sensitivity of traditional agricultural production to changes in temperature and rainfall patterns. This sensitivity could affect mostly small producers in the region due to less technological and infrastructure development (REGATTA 2014; IPCC 2014b).

In this regard, it should be borne in mind that in recent years extreme weather events (such as prolonged droughts, floods, frosts, heat waves, etc.) have highlighted the high vulnerability of the region's production system to such situations, with significant losses in production. By way of example: (a) Argentina lost 2.8% of its GDP with the drought in 2008; (b) Paraguay lost around USD 1.25 billion in the sector, as a consequence of four events: two of drought in the years 2004–2005 and 2007–2008, and two of excess rainfall in the years 2000–2001 and 2011–2012. The main problems are associated with social and productive variables such as job losses and precarious employment, soil degradation, unequal land distribution, the effects of climate variability, water availability and access, the degree of use of irrigation systems, and agricultural insurance coverage.[9] An analysis of these variables shows that the vulnerability of the region's production system is at a level similar to the Latin American average. However, it is noted that in some variables there is a greater manifestation, such as: the fragility and erosion of soils and the loss of productivity

[9] In the case of agricultural insurance, whose use could reduce the impact of events, its use is not yet widespread, especially among small producers. Thus, for example, agricultural insurance coverage in Argentina is around 60% nationwide, although it currently covers mainly hail and the inclusion of drought has only just begun. In Paraguay, coverage is limited to the business sector, and only about 8% of the area under soybean and some maize cultivation is covered, although there are proposals at the central government level to implement it within the band that does not have access to the insurance tool. In Bolivia, the process is also being initiated at the central government level with the same objectives (REGATTA 2014).

due to poor management, as well as genetic deterioration due to invasion and introduction of invasive species. Another factor that increases vulnerability to climatic factors is the low level of technology in the production system. For example, the percentage of use of irrigation systems in the region's agriculture reaches very low levels, and they are generally used for rice, vegetables, pastures and, to a lesser extent, sugar cane.

In recent years, various studies have been carried out in Argentina (Argentina 2007, 2015; ECLAC 2014a), Bolivia (Bolivia 2009; ECLAC 2007), and Paraguay (Paraguay 2011; ECLAC 2014b) on the potential impacts of climate change on agricultural production and its importance from an economic point of view. However, these studies were not necessarily focused on the Gran Chaco region. The studies mentioned for the case of Argentina, for example, were linked to the changes that could occur in the yields of the country's main crops (soybeans, wheat, corn) but mainly in the core area of Argentine agriculture (Pampean Plain).

In the case of Bolivia, its Second National Communication on Climate Change (Bolivia 2009) proposes the potential economic impacts of climate change on the agricultural, forestry, and livestock sectors, as key areas of the Bolivian economy, based on a scenario analysis. The study shows that there is a negative correlation (inversely proportional) between the growth rate of agricultural GDP and the occurrence of the El Niño phenomenon, while the La Niña phenomenon is not a relevant factor in determining the growth of the sector's GDP. When there were strong El Niño events (1982–1983, 1991–1993, 1997–1998, 2006–2007), there was a sharp fall in agricultural GDP. The reduction in potential areas ranges from 2.4 to 84% of the affected surface area, which implies a drop in production and the loss of important forest areas and also grazing areas. According to ECLAC (2007), the economic impact of El Niño 2006–2007 on Bolivia's agricultural sector was 79.6 million USD (14% of agricultural GDP and approximately 1% of national GDP), with the main losses being in soybeans (44 million USD) and rice (17 million USD). In this case, the loss is 9% of the total cultivated area and 83% of the total affected area. Two-thirds of the lost area is soya and rice. As for the livestock sector, in this case the direct relationship is established with the La Niña phenomenon (e.g., 1999–2001) in which a reduction in meat production was observed. It is estimated that during the La Niña phenomenon, losses in the livestock sector can triple compared to the results obtained during El Niño (ECLAC 2007).

Paraguay's Second National Communication on Climate Change (Paraguay 2011) also analyzes the potential impacts of climate change on the agricultural sector. In this case, industrial crops were taken: soybean (grown by large and medium producers, the country's main export item), cotton (produced by small and medium farmers, the main source of income for small producers), sesame and sugar cane (food and potential bioethanol production) and popular consumption crops: corn, cassava, and beans. They were selected for the area occupied in terms of their participation in income and for their importance in feeding the population. Bovine meat and milk production were also taken into account because of their impact on the population's diet and on the generation of income from exports. Historical trends show:

- Falls in yields for cotton in the period 1990–2006 (from 1400 to 850 kg/ha), to a lesser extent for sugar cane, in the same period (from 47.5 to 47.38 ton/ha) and a fall of about 500 kg/ha in sesame yields (from over 1100 to about 600 kg/ha).
- A small increase in maize yields; given an improvement in the technology applied (from 1890 to 2500 kg/ha) and a significant increase in bean yields.
- Almost unchanged yields for manioc (around 11,700 kg/ha) and soya (in this case for the period 1992–2006).
- In terms of predicting changes in yields, the results are as follows.
- A continuation of the declining trend with yields falling increasingly over the years (halving from current yields by 2020 to 429.67 kg/ha); while for soya the outlook is not very encouraging, as yields would fall sharply in a relatively short time.[10]
- For sugar cane, on the other hand, projections show a constant trend between 47 and 45 ton/ha. In the period 2007/2020, the yield is going to be slightly below 47 ton/ha, to increase slightly after a range of 47–48 ton/ha.
- Maize yield would continue to increase from 2504 to 3000 kg/ha by 2020 and to 4262 kg/ha by 2050. Yields would also increase for manioc, from 17,611 to 29,616 kg/ha in 2020 and continue to grow toward 2050. And the same would be true for bean yields, which would increase to 935.76 kg/ha by 2020. Similarly, sesame yields are expected to increase from 779 to 2155 kg/ha by 2020.
- Meanwhile, for livestock, a positive evolution is expected, expecting around 51 million heads by 2020.

Beyond these efforts made by the countries that have the largest areas of their territories making up the Gran Chaco, only in REGATTA (2014) is an analysis made that corresponds exclusively to the region in question. In this sense, the impacts on the agricultural sector are analyzed based on the relationship between crop yields and the behavior of climatic variables.

The results of the yields of the various crops are compared with the trends in total annual rainfall for the period 1961–2040 and the trend in total annual temperature for the same period, by department, for the IPCC SRES A2 scenario. Once the variation in the yields of each crop by department was determined, the sensitivity of the sector was calculated, which consists of the weighted sum of each of the items according to the value determined by the five prioritization criteria (cultivated area, production value, number of producers, number of unit of analysis, participation in the basic food basket). Thus, the areas in which the crops tend to have a high, medium, or low sensitivity were obtained. In the same way, the items are classified according to their yield, in items of high, medium, and low sensitivity. The results can be seen in Fig. 3.2 and Map 3.14.

This sensitivity analysis was also carried out for meat and milk production (REGATTA 2014) and formed part of the determination of vulnerability based on the expression (3.1):

---

[10] This behavior was corroborated, while in reality soybean yields at least did not increase, according to CAPECO (2016).

Fig. 3.2 Agricultural yields by crop. *Source* Adapted from REGATTA (2014)

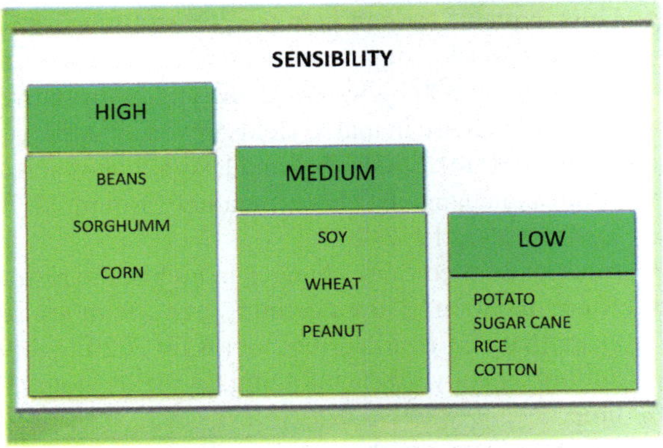

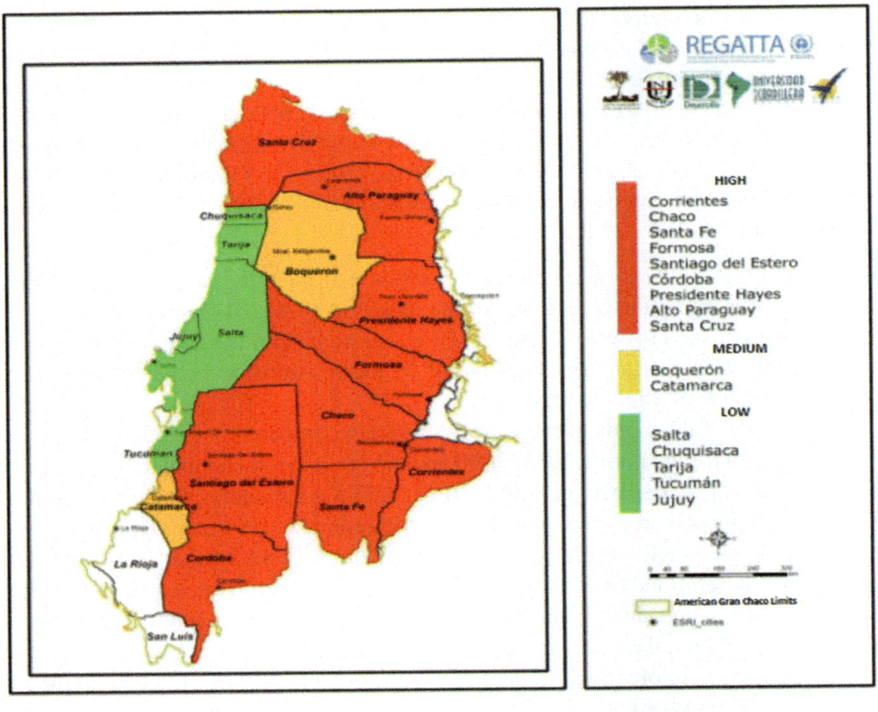

Map 3.14 Sensitivity for city/department. *Source* Adapted from REGATTA (2014)

$$\text{Vulnerability} = (\text{Sensitivity} + \text{Exposure}) - \text{Adaptive Capacity}. \quad (3.1)$$

Vulnerability and adaptive capacity are multidimensional, complex and not directly observable, making them difficult to assess. In this case, in order to combine the different variables into a single value, indices were constructed with the sectoral

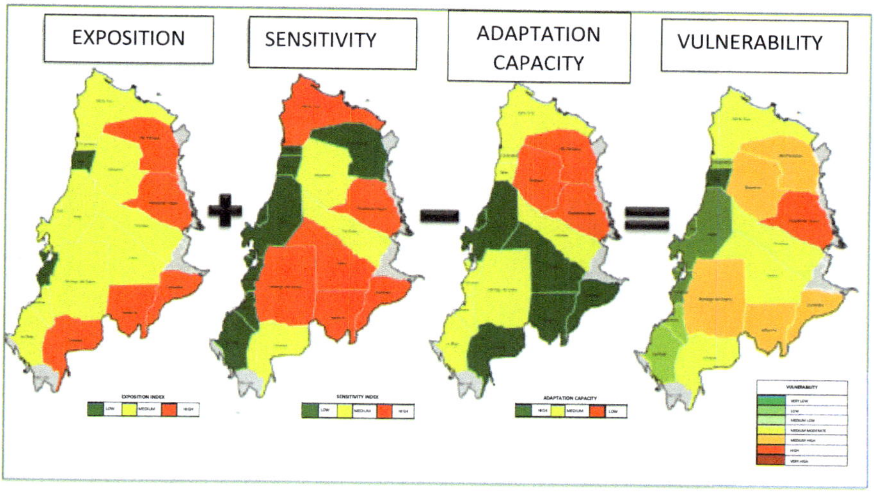

Map 3.15 Vulnerability for decade 2011–2020. *Source* Adapted from REGATTA (2014)

variables.[11] Finally, from this information, Maps 3.15, 3.16, and 3.17, of exposure, sensitivity and adaptive capacity were drawn up for the period of analysis, grouped into decades, from 2011 to 2020; 2021 to 2030 and 2031 to 2040, respectively.

Map 3.15, which corresponds to the first decade (2011–2020), shows that the Northeast and Central areas are the most vulnerable in the region. This is explained by the fact that it is the area that presents the highest climatic exposure (especially high temperatures and extreme events), despite the fact that some units of analysis present high adaptation capacity (Santa Cruz) or low sensitivity (Alto Paraguay). In this decade, exposure predominates over the final result of high vulnerability. It is worth mentioning the Paraguayan departments of Presidente Hayes, Alto Paraguay and Boquerón, areas with high vulnerability. These are attributed, in addition to the high exposure and sensitivity, to their low adaptive capacity. In the Central-South zone of the region, vulnerability remains in the medium range. This is due to the fact that despite having medium level exposure and high adaptive capacity, sensitivity is also high. In contrast, the Western zone, following the line of the Andes Mountains, is the area that presents the least vulnerability to climatic conditions. This behavior is explained by having a medium level of exposure, with low sensitivity in production and a high capacity to adapt. This is why the vulnerability remains in low ranges (REGATTA 2014).

[11] The exposure index is constructed with three variables: the deviation of the amount of precipitation of the period under consideration from the base period, the deviation of the temperature from the base period, and the number of extreme events. The sensitivity index was obtained by adding the water scarcity index to the agricultural and livestock sensitivity indexes. Adaptive capacity is related to a society's means of mitigating potential damage from climate change. This indicator includes elements such as: the Human Development Index (HDI), the value of production per producer measured in US dollars, the area cultivated under irrigation, the silo and storage infrastructure, etc.

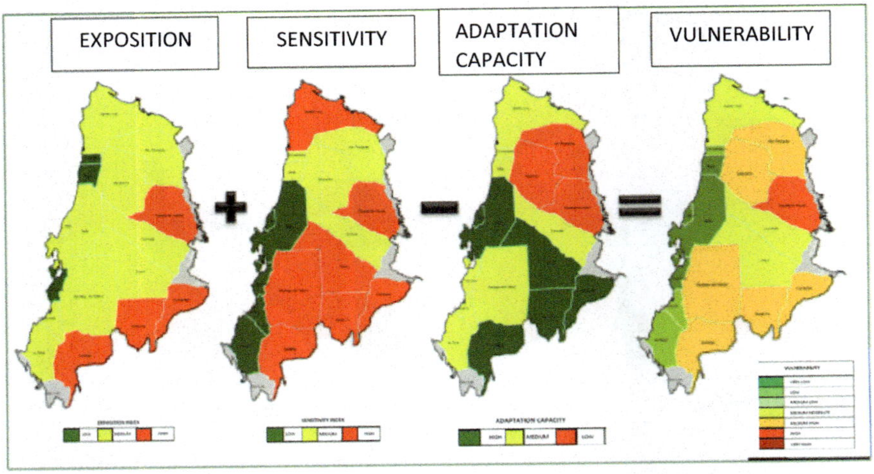

Map 3.16 Vulnerability for decade 2021–2030. *Source* Adapted from REGATTA (2014)

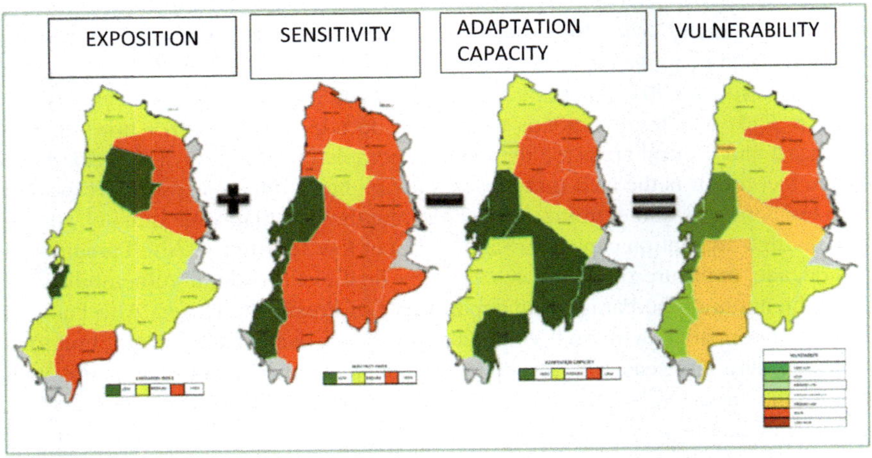

Map 3.17 Vulnerability for decade 2031–2040. *Source* Adapted from REGATTA (2014)

Map 3.16 shows the components of vulnerability for the decade 2021–2030. The behavior is similar to that of the previous decade, except for Córdoba and Tarija, which increase their vulnerability. This increase in vulnerability is attributed to the greater sensitivity of agricultural items, especially agricultural crops (REGATTA 2014).

Map 3.17 presents the components of vulnerability in the last decade analyzed (2031–2040). It can be seen that exposure is increasing throughout the region, with temperature increases and extreme events toward the end of the analysis period.

These conditions also affect sensitivity, especially agricultural production, as well as a decrease in water availability. The high vulnerability of the Paraguayan Chaco, in addition to the high exposure and sensitivity, is attributed to the low adaptation capacity, mainly due to deficiencies in the management of its resources and institutions (REGATTA 2014).[12]

In terms of vulnerability, one fact that cannot be ignored is the presence of extremely vulnerable communities in the region. For example, in the Paraguayan Chaco, there are areas where the vast majority of the population bases its food security on hunting and gathering in the forest.

3.6 The Semi-arid Region in Brazilian Northeast (the Sertão)

In Brazil, the areas prone to desertification exceed 1.3 million km^2, which is equivalent to 16% of the country's territory, as shown in Map 3.18. These areas concentrate around 34 million inhabitants, approximately 17% of Brazil's population, making it the most populated dry region in the world, home to 1488 municipalities (Brazil 2016). Brazil is classified as a medium risk country in the recently elaborated IVCC (CAF 2014). However, the main focus is on the semi-arid zone of the Brazilian Northeast.

Map 3.19 shows the geographical location of the Brazilian Sertão Region which includes the states of Ceará, Rio Grande do Norte, Paraíba, Pernambuco, Alagoas, Sergipe, Bahia, Minas Gerais and Piauí, in the Northeast of Brazil. This semi-arid area covers 1 million km^2 and is populated by over 20 million people. This region of the Brazilian semi-arid is expected to be one of the areas most affected by climate change in Brazil. However, this region is also the most vulnerable within the regions of Brazil due to its low indexes of social and economic development, bringing together the worst social indicators of the country. A large part of its population is engaged in rainfed agricultural activities (which pose high risks to agricultural activity in the presence of scarce and variable rainfall), with a very low degree of technification and high dependence on the availability of natural resources, mainly water (IPCC 2014a, b, c; Angelotti et al. 2015; Brazil 2004, 2007, 2010, 2016).

This semi-arid region of Brazil is characterized by high evapotranspiration, prolonged periods of drought, shallow soils, high salinity, low fertility, and reduced water retention capacity. All these factors limit its productive potential. In this way, the process of desertification is intensified by poverty and vice versa. The most alarming social indicators in Brazil are in this region (Brazil 2016).

The consequence of this combination of factors is slow but persistent environmental degradation (see Map 3.20). In this sense, erosion is the most serious case of

[12] It is recalled that in CAF (2014), both Paraguay and Bolivia have extreme vulnerability indexes, while Argentina has a medium index. In South America, Paraguay and Bolivia have the highest vulnerability risks as measured by the CAF (2014).

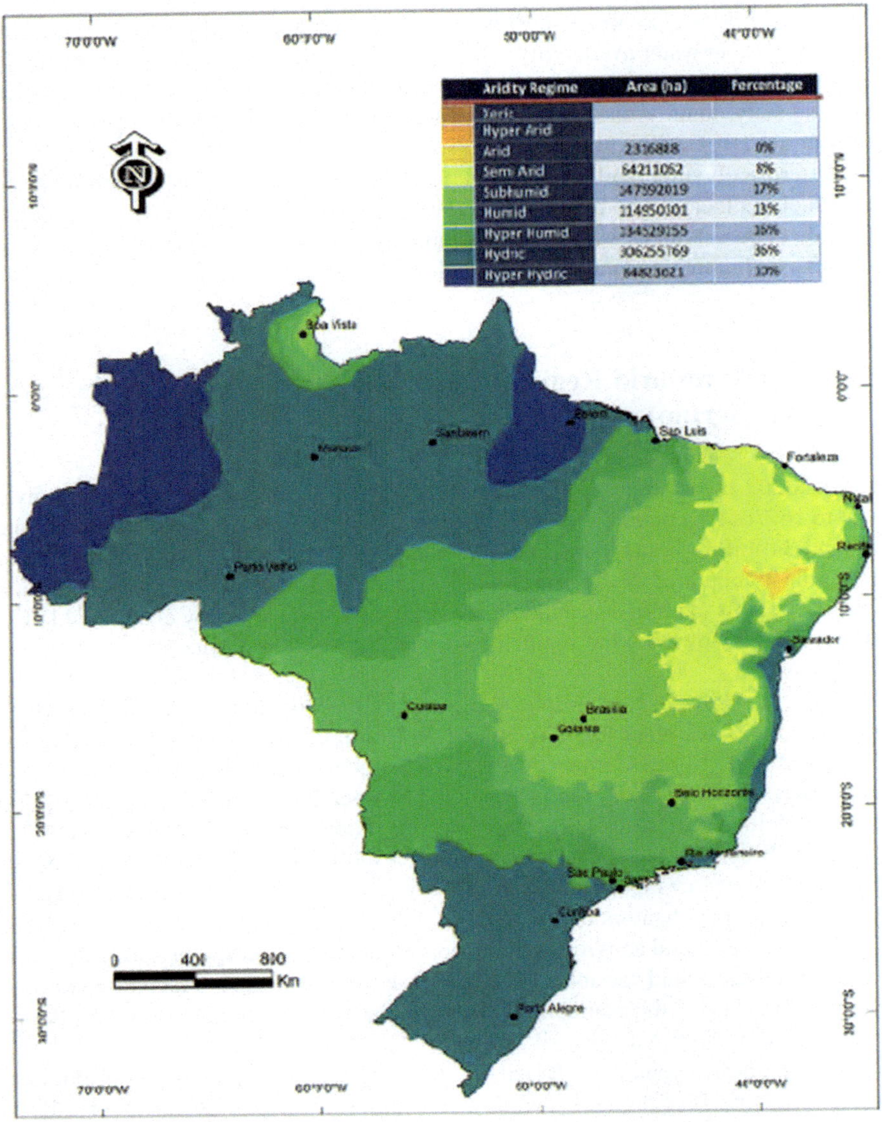

Map 3.18 Arid and semi-arid regions of Brazil according to the Atlas of arid zones in Latin America and the Caribbean. *Source* UNESCO/PHI-LAC (2010)

Map 3.19 The Sertão Region. *Source* https://www.ibge.gov.br

environmental degradation in the semi-arid NE of Brazil, given its irreversibility. To this must be added the large extension of shallow soils, the occurrence of torrential rains, and the development of agriculture in areas with steep slopes and without any preventive measures (Brazil 2016; IPCC 2014c).

Desertification in this region is caused by a complex interaction between physical, biological, social, economic, political, and cultural factors, in a vicious circle of (a) deforestation, (b) soil degradation, (c) reduction of agricultural production and income, and (d) deterioration of social conditions. Therefore, the mismanagement of dry forest (*Caatinga*) resources, and sometimes of the *Cerrado*, cannot be ignored. When it comes to explaining the problem of desertification in Brazil Agricultural practices without proper soil management, the indiscriminate use of irrigation systems (with the consequences of salinization) and overgrazing (in cases of extensive exploitation) compromise the regeneration of species. This lack, associated with deforestation, causes erosion and soil depletion (Brazil 2007, 2016; IPCC 2014c).

As regards future scenarios, these point to a continuation of the trend of deepening drought in NE Brazil, leading to greater aridity in the semi-arid region by the end of the twenty-first century, which could directly influence the characteristics and distribution of vegetation. The higher the air temperature, the greater the tendency for the semi-arid water deficit to increase, considerably affecting human and animal water

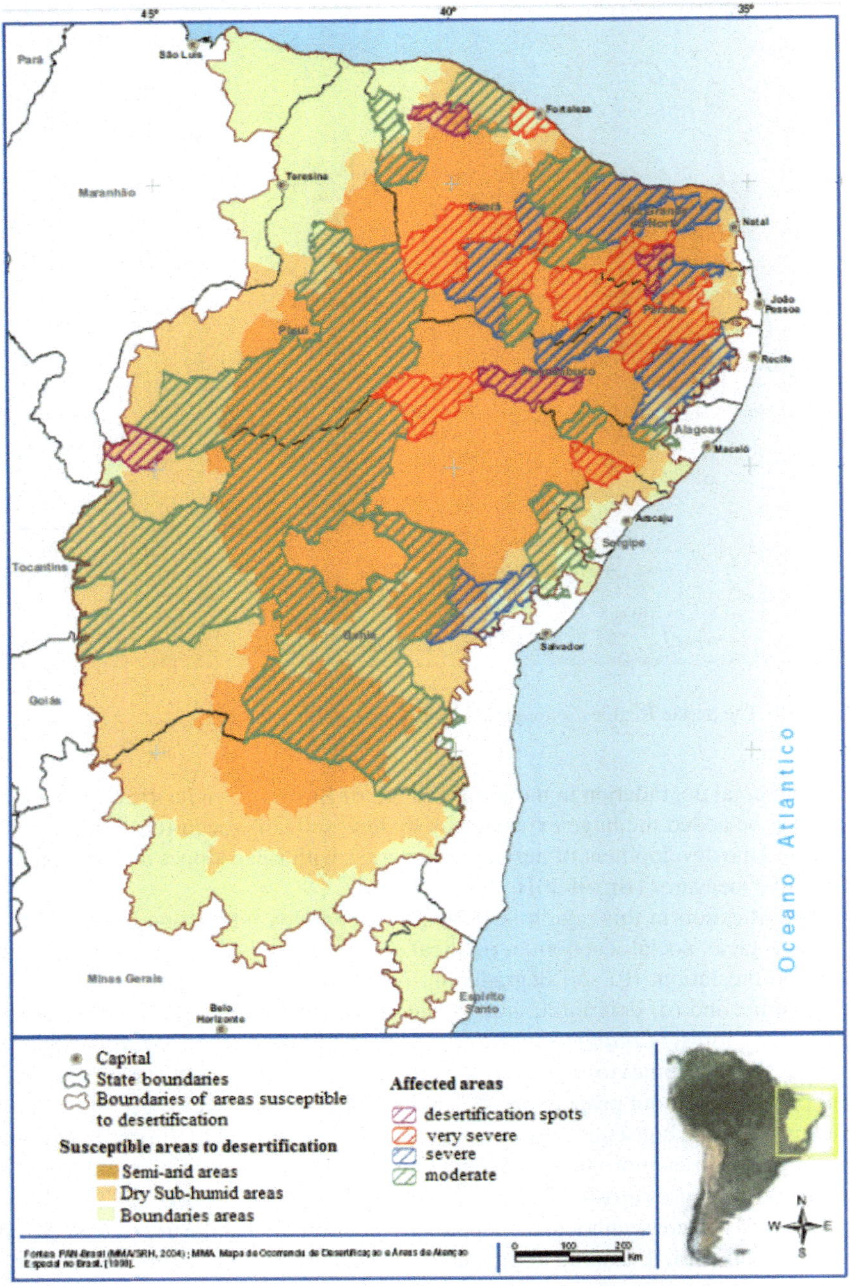

Map 3.20 Affected areas and desertification spots in the northeast region of Brazil *Source* Brazil (2007)

consumption, as well as activities that depend on rainfall (IPCC 2014b; Angelotti et al. 2015).

Thus, impacts due to temperature increases and precipitation anomalies (long periods of combined drought and heavy rainfall) may significantly affect crop production, water resources, irrigation management, biodiversity and desertification processes. Thus, potential negative impacts on water resources and their consequences on rainfed agriculture could compromise the livelihoods of the region's population (Brazil 2004, 2010; IPCC 2014b; Angelotti et al. 2015).

With these forecasts, there is a strong demand for research to assess the effects of climate change in the Brazilian semi-arid. Research is proposed to find resilient ways of adapting to the climate and to generate knowledge to apply technologies, processes, and actions to reduce or manage climate risks in the region. In this sense, it is of crucial importance to raise awareness about the need for a comprehensive vision of the problem of adaptation (Angelotti et al. 2015).

Among the adaptation measures being considered are the following (Angelotti et al. 2015):

- Adaptation actions supported by the integration of various biological diversification measures to combat desertification.
- Ecosystem-based adaptation actions (establishment of protected areas), as well as community conservation and management of natural areas.
- Adaptation actions in agriculture, through rainwater harvesting, efficient use of irrigation water, conservation of soil moisture through the use of dead plant cover, genetic improvement with selection of materials resistant to drought and high temperatures, application of multiple crop (or polyculture) systems, use of local genetic diversity, among other measures.
- Adaptation actions related to the water issue, through the application of management techniques to increase water use efficiency: the use of underground storage, emergency irrigation, and drip irrigation.
- Adaptation actions in agricultural techniques used to minimize the impacts of agriculture, such as organic fertilization, as well as the use of stubble both to increase the capacity to retain soil moisture and to achieve a reduction in evapotranspiration losses. The use of spontaneous vegetation is also recommended as an alternative to dead cover (scarce, in the semi-arid) that can increase water use efficiency.
- In the case of monocultures, temporary measures such as altering sowing dates, the introduction of new cultivars, and the use of irrigation can be applied.
- An alternative is crop diversification, seeking to ensure that heterogeneity on the scale of the landscape can effectively help to increase agricultural production.
- The development of silvopastoral and agroforestry systems. These systems, in the semi-arid, have to contemplate the integration of goats and sheep with commercial systems of permanent forest species and in dry farming systems. As an additional effect, it is verified that the agroforestry system reduces the loss of soil, water and nutrients and, consequently, delays the erosion process; being also useful to accumulate carbon. These agrosilvopastoral systems can consist of areas with

the introduction or maintenance of the tree component (native or exotic) with cultivated pastures adapted to the semi-arid and systems with the use of fodder palms, corn, grasses and fodder legumes. Or these agrosilvopastoral systems can also be developed through the integration of tree species (native or exotic) with crops adapted to the semi-arid such as manioc, sorghum, or cowpea.

- Opportunities for the development of the semi-arid can be explored by applying solutions linked to the payment of environmental services and to postulates related to the green economy, in order to try to minimize the impacts of economic activities, taking into account that it is the region most vulnerable not only to climate change but desertification as well.

Considering that the concern for conservation as an adaptation measure is global, different adaptive strategies will have to be applied to overcome the expected negative impacts of climate change on existing agricultural systems in the semi-arid. In this sense, it is very important to pay attention to research. In the case of Brazil, for example, the relevance of a series of research projects, participatory management and use of agrobiodiversity by communities in the area should be highlighted. From these, an important diversity of cultivated and native species was identified; for example, 59 varieties of cassava and 55 varieties of beans. These studies not only contribute significantly to the food security of the population living in the semi-arid, but are also of fundamental importance both for this region and for other states in Brazil, where future climate scenarios foresee similar situations to those in the semi-arid (Angelotti et al. 2015).

3.7 The Venezuelan Coastal Zone

According to the study carried out by the CAF on vulnerability to climate change, Venezuela has an IVCC that implies a high risk (CAF 2014). The country's coastal area is the one with the highest vulnerability indexes. This situation can be seen on both Maps 3.21 and 3.22.

Various works (IPCC 2014b; Venezuela 2005; Gabaldón 2008; Díaz 2001; Martelo 2004; Puche et al. 2008) present the main expected impacts of climate change on Venezuela. Some of these impacts are estimated on the basis of the previous emissions scenarios prepared by the IPCC in its 4AR for 2007 (mainly SRES A2 and B1, the former more "pessimistic" and the latter more "optimistic" in terms of results on temperature and rainfall toward the end of the twenty-first century) while others arise as a result of the application of the new IPCC scenarios (RCP) from its latest 5AR report. However, the results (many of which are contained in Venezuela's First National Communication to the United Nations Framework Convention on Climate Change and others obtained subsequently) are consistent with the trends they show. Some of these can be listed below.

Lower rainfall (and consequently drier weather) is expected from June to February during part of the rainy season (for various climate models used) throughout the

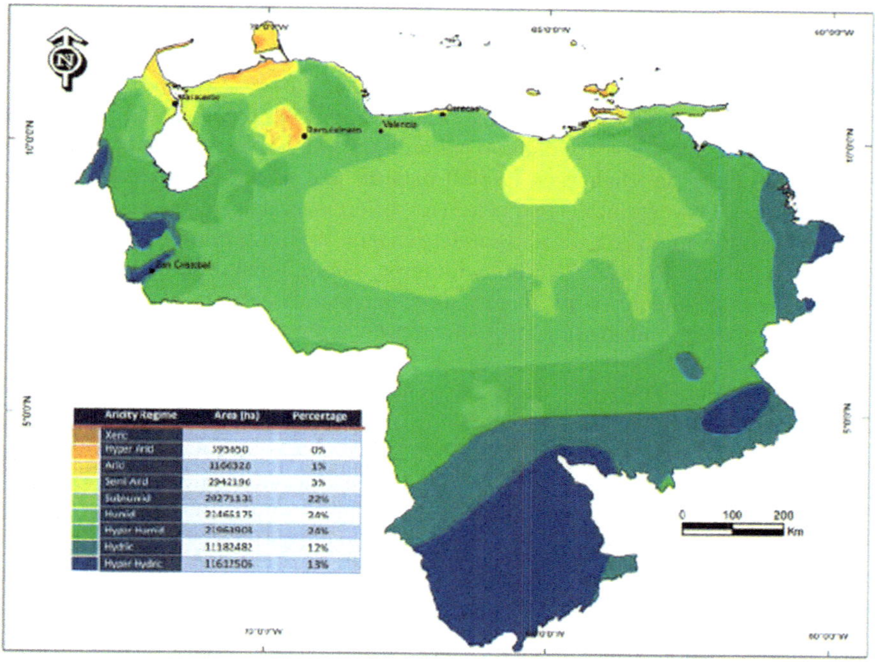

Map 3.21 Arid and semi-arid regions of Venezuela according to the Atlas of arid zones in Latin America and the Caribbean. *Source* UNESCO/PHI-LAC (2010)

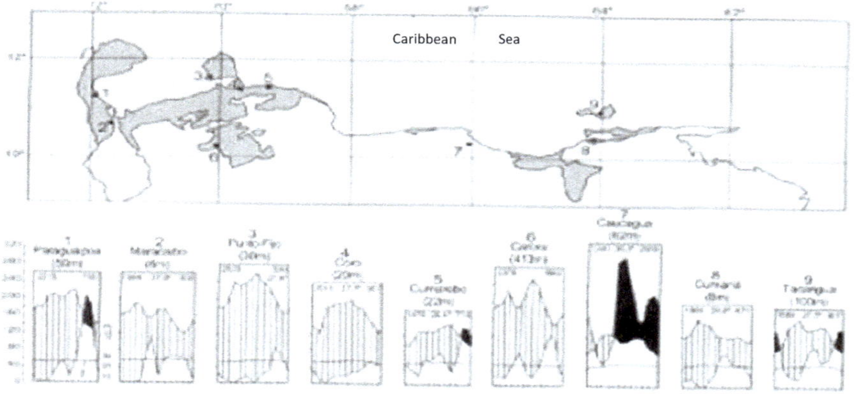

Map 3.22 Semi-arid region of Venezuelan coastal zone. *Source* Adapted from Díaz (2001)

central strip of the country, from the Andes to the Eastern Llanos and north of the state of Bolivar (Guyana area) and the same is expected for the extreme south of the state of Bolivar and the extreme north of the state of Zulia (Maracaibo).

Both increases and decreases in precipitation and runoff associated with higher and more frequent extreme (maximum or minimum) values are expected, aggravating the effects of climate change on rainfall patterns and its social and environmental consequences. Lower rainfall is expected (for different models) in the south of Bolivar State. The results indicate a warming and a trend toward lower rainfall in the future (in the different scenarios analyzed) but with important regional differences. The south seems to be more affected while there is more uncertainty in the mountainous regions (the Andes and the central north). A temperature increase is expected throughout the territory.

The arid climate currently covers 2% of the country's surface area, the semi-arid climate 11%, and the subhumid climate 26%. The various models used estimate that this percentage of the territory, which is under climates considered critical by the United Nations Convention to Combat Desertification, may raise from 39 to over 47% of the country's surface area by 2060. The areas with aridity (with rainfall of 500–600 mm per year) are located mainly in the states of Lara, Zulia, Falcón, and a large part of the coastal area. The increase in arid, semi-arid, and subhumid climatic zones will increase the vulnerability of soils to desertification and degradation.

Venezuela's agricultural sector is structurally weak. Most agricultural activities are carried out under rainfed conditions. Only a small part is carried out under irrigation. Thus, there is an intensive use of both soil and water, and thus agriculture is highly dependent on technological inputs that cause or may cause problems both by deterioration of natural resources and by environmental pollution.

There are 7,950,000 ha that are destined for vegetable agriculture; 9,280,000 ha for mixed exploitation; 18,420,000 ha for livestock and 19,460,000 ha for forestry use. Vegetable agriculture: Guárico, Portuguesa, and Barinas. Cultivated pastures: Zulia, Falcon, and Barinas. Typical livestock system of the Venezuelan plains, low productivity, high mortality. The main crops are: corn, rice, coffee, sorghum, sugar cane, cocoa, banana, yucca, sesame. See Maps 3.23, 3.24, 3.25 and 3.26.

The results show that the number of wet months changes (due to the variation in precipitation and evapotranspiration and the general rise in average temperatures). This can significantly alter the spatial–temporal distribution in the areas of highest agricultural production in the country, with consequences in the decrease of yields of some items, such as maize. The increase in night temperatures leads to a decrease in the net accumulation of dry matter and therefore a reinforcement of the decrease in yields. This is true for the country's main rainfed crops (Venezuela 2005; Puche et al. 2008).

The expected changes in the yields of the main crops, based on the modifications in the climatic variables, were estimated in four points taken as case studies: El Tigre, Santa Cruz, Turén, and Calabozo, for maize, rice and caraota (black beans) crops. See Map 3.27.

The changes observed in yields are as follows: a reduction in maize yields from 2.3 to 4.4% by 2020 and from 6.2 to 12% by 2060. In the case of rice, the drop in

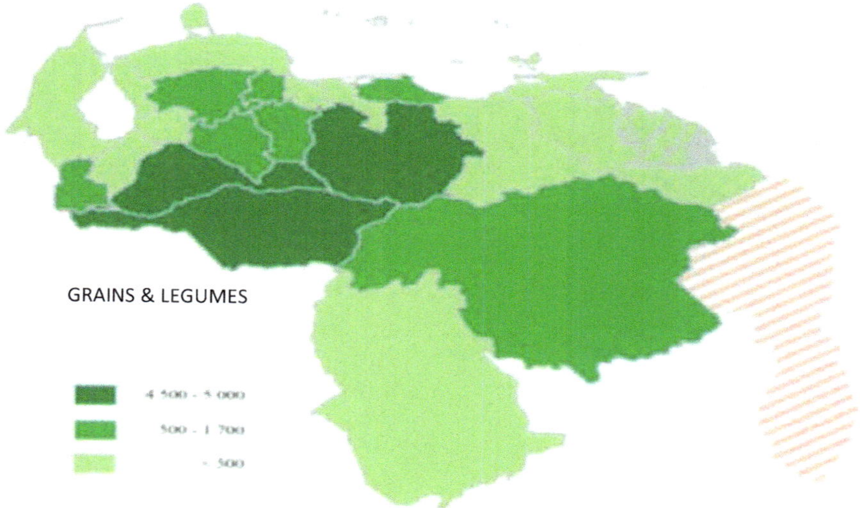

Map 3.23 Grain and legume production (in tons). *Source* Adapted from Venezuela (2005)

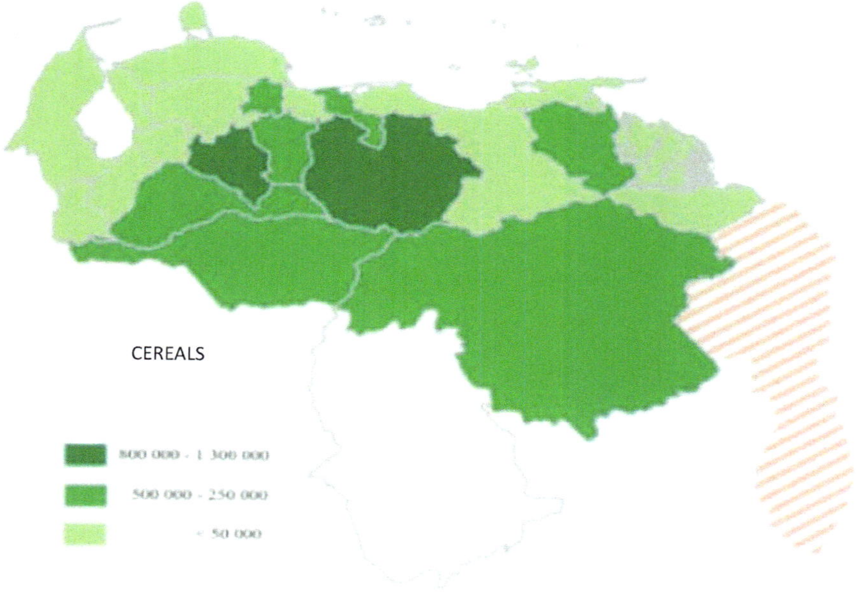

Map 3.24 Cereal production (in tons). *Source* Adapted from Venezuela (2005)

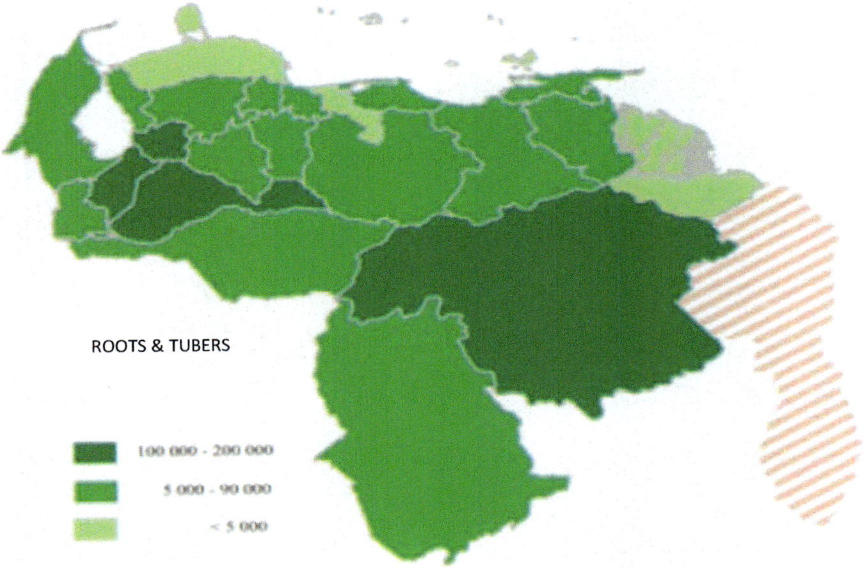

Map 3.25 Root and tuber production (in tons). *Source* Adapted from Venezuela (2005)

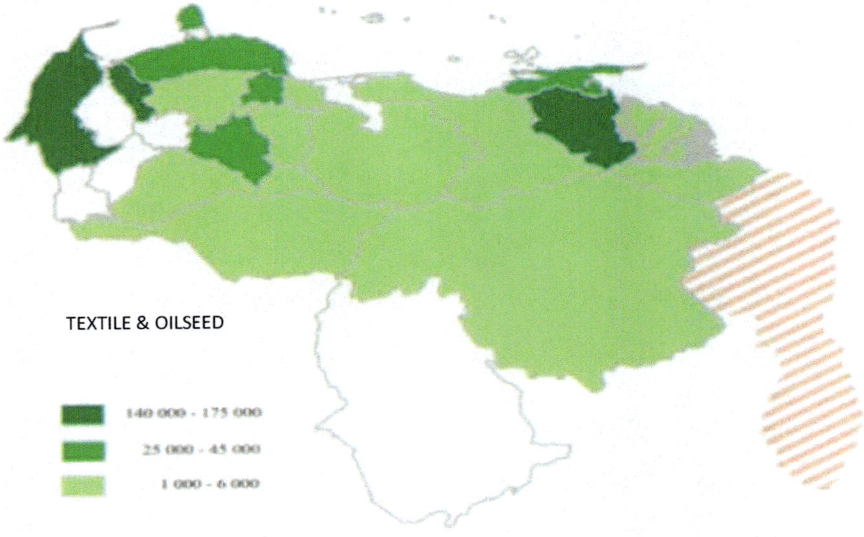

Map 3.26 Textile and oilseed production (in tons). *Source* Adapted from Venezuela (2005)

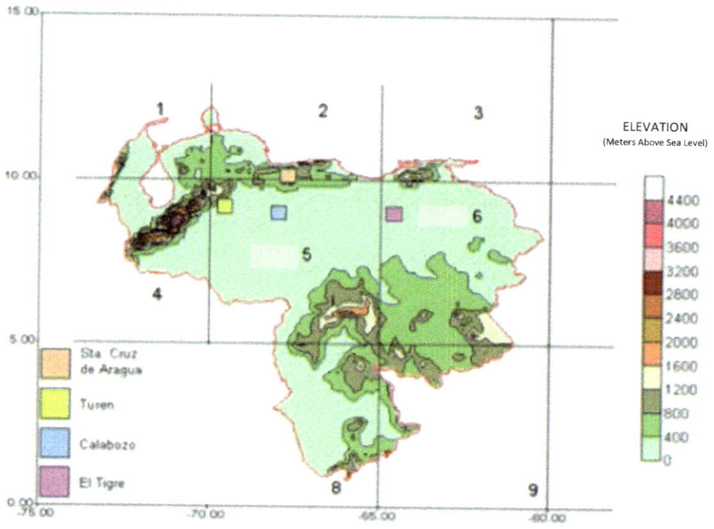

Map 3.27 Characteristics of the sites taken as a case study to determine the expected impacts of climate change on the yields of Venezuela's main crops. *Source* Adapted from Venezuela (2005). *Note* Santa Cruz de Aragua = Central Zone; Turen = Western Llanos; Calabozo = Central Llanos; El Tigre = Eastern Llanos

yields is from 3.1 to 4.4% by 2020 and from 7.6 to 11.8% by 2060. And in the case of beans, the reductions range from 2.2 to 13.4% by 2020 and from 8.7 to 43.2% by 2060 (Puche et al. 2008).

The increase in the minimum temperature seems to be the main factor in the drop in yields, even in the locations where irrigated crops were simulated (Venezuela 2005). A slight shortening of the crop cycle and reduced water availability is observed.

Flooding and increased erosion of coastal landscapes are expected. Mainly on the eastern shore of Lake Maracaibo, on the coast of Guajira and in departments such as Falcón, Carabobo, and Sucre, among others.

A deepening of the degradation of agricultural soils and a loss of fertility is also expected, mainly in the cultivated areas on high slopes in the areas of the Andes, the coast and the arid and semi-arid zones with scarce plant cover such as Falcón, Lara, and Zulia. An intensification of the processes underway in the central and eastern plains, where extensive areas of degraded soils are found, is also expected (Table 3.3).

3.8 The Semi-arid Belt of Central America

Central American countries exhibit relatively high levels of exposure to climate change and, at the same time, present the most extreme risks of vulnerability as measured by the Climate Change Vulnerability Index (IVCC) developed by the

Table 3.3 Expected impacts of climate change on water regimes by region and by crop

States affected		Impact description	Items affected	Possible impacts to current uses
Zone of impact: Oriente (The East)				
Sucre		An increase in the affected area is observed, in which the range of precipitation will decrease from 1200–1600 to 800–1200 mm/year, extending to areas of agricultural importance	• Cocoa, sugar cane, coffee, coconut, fruit trees (mango, cambur), and yucca • Solanaceous vegetables • Subsistence agriculture	
North of Moragas	Caripe and Caripito		• Coffee, orange, cocoa • Leaf vegetables • Subsistence agriculture	Conflict could be created with cities over water, irrigation systems could be affected by water deficiency
	Maturín (Guarapiche Valley)	Marginal zone from the water point of view. Rainfall in the affected area will decrease from 800–1200 to 400–800 mm/year	• African palm, annual crops in valley areas • Tobacco under irrigation • Irrigated floor vegetables • Mangoes and limes • Subsistence agriculture	
South of Moragas		In this area there is a decrease in rainfall from 800–1200 to 400–800 mm/year	• Cassava (yucca), pine (500.00 ha). Introduced pastures and grasslands (which are 20%) • Subsistence agriculture	Medium/high technology cassava plantations could be affected by the lack of water
Anzoátegui		The area with the least amount of rainfall around Anaco expands toward the Mesas in El Tigre. The area will increase from 800–1200 to 400–800 mm/year. In the very small area to the north of El Tigre (Aguasay, Anaco)	• Grasslands, seeds, fruit trees, creepers (melon, sideburns) • Livestock farming • Cassava • Subsistence agriculture	The demand for crop irrigation is going to increase

(continued)

Table 3.3 (continued)

States affected	Impact description	Items affected	Possible impacts to current uses
Delta	The decrease is very small from 1600–2000 to 1200–1600 mm/year	• Rice, pastures • Bananas in small non-commercial areas • Subsistence agriculture	The crops depend more on the water level of the pipes: here the problem would be the sea water, due to the influence of the saline curve, which could affect the area
Bolívar	The northern zone will be affected by a decrease in rainfall from 1200–1600 to 800–1200 mm/year. The main production zone, La Paragua. There are no major changes to the south in either model, zones of 2600 mm go to 1800–2000 mm/year	• Corn, fruit trees, small eucalyptus plantations (North of Caroni) • Corn in La Paragua • Tomato • Yucca and yam • Subsistence agriculture	No impact described
Zone of impact: Unare Basin			
	It appears that the area with the least precipitation will expand from 1200–1600 to 800–1200 mm/year. The drought in the Unare Basin will be accentuated. It is one of the areas with a high number of reservoirs (approximately 15)	• Dual purpose livestock farming • Pasture land • Subsistence agriculture	The demand for water from grasslands could be affected
Zone of impact: Central Occidente (Central West)			

(continued)

Table 3.3 (continued)

States affected	Impact description	Items affected	Possible impacts to current uses
Aragua, Carabobo, Miranda, North of Guárico and Vargas	It is an area of 800–1200 mm/year that covers the coastal area of Vargas, Aragua, and Carabobo (Puerto Cabello). The area where there was rainfall of 1600–2000 mm/year will disappear and dominate the range of 800–1200 mm/year, according to the UKTR model. It will expand to the North of Guárico (San Juan de los Morros and South of Aragua). This area will be affected by the expansion toward the South, reaching Carabobo and by the West it joins the driest area of the Lara State affecting the Yaracuy State	• Cocoa (under irrigation on the coast) • Banana • Cambures cane, seeds (cereals), fruit trees (in Aragua valleys) • Pepper, tobacco, tomato, and vegetables (in the Guárico river valley) • Southern Aragua and Northern Guárico grasslands • Subsistence agriculture • Poultry farms (Carabobo)	This area does not change its regime, what it does is expand. So the area under irrigation could be shared with tourism. They will have to look for changes in their (more efficient) irrigation systems Chicken production in Carabobo can be affected by the increase in temperature In Guárico water supplies from the Camatagua reservoirs and Guárico itself will be affected
South of Carabobo	Areas with precipitation of 1200–1600 go to 1200–800 mm/year	• Citrus fruit • Sugar cane • Vegetables • Subsistence agriculture	The area will be affected by droughts; the area with the lowest rainfall is expanding
Yaracuy (Bajo Yaracuy: San Felipe-Municipality. Veroes-Manuel Monje)	Zone of 1200–1600 goes to 800–1200 mm/year	• Sugar cane (Central Santa Clara) • Musaceous • Citrus fruit • Grasslands • Subsistence agriculture	

(continued)

Table 3.3 (continued)

States affected	Impact description	Items affected	Possible impacts to current uses
Yaracuy Middle (Municipality Bruzual, José Antonio Paéz, Peña, Urachiche)	Rainfall from 1200 to 1800 (Chivicoa-Yaritagua) increases to 900–1200 mm/year	• Corn and now also sorghum • Sugar cane (Turbio River) • Secondary: quinchoncho beans, caraota	The planting of crops with higher water demand, such as sugar cane, would decrease. Sugar cane is at the limit of temperature. The planting of crops that are less demanding in terms of water, such as sorghum, beans, and quinchoncho, would increase
Yaracuy High Valleys (Municipality of Nirgua)	Precipitation zone of 600–900 (Municipality. Peña which includes La Piedra-Yaritagua). Increases from 450 to 600 mm/year	• Citrus • Fruit trees (avocado) • Secondary: mushrooms, legumes, corn	The planting of less water-demanding crops such as sorghum, beans, and quinchoncho will be increased
Zone of impact: Occidente (The West)			
Lake Maracaibo Basin (East Coast of the Lake)	According to the CCC-EQ model, there is a zone that goes from 1200–1600 to 800–1200 mm/year between Ciudad Ojeda and the Moratán Delta and parallel to the Trujillo mountain range, leaving only a small area between Mene Grande and the north of the Motatán Delta	• Grassland • Corn • Sugar cane • Bananas, Cambures • Yucca (Trujillo)	In this area the irrigation system of the Cenizo River could be affected, as well as the production of bananas and corn in the Motatán Delta

(continued)

Table 3.3 (continued)

States affected	Impact description	Items affected	Possible impacts to current uses
Lake Maracaibo Basin (South of Lake Maracaibo)	The CCC-EQ model indicates in the area of the Pantanos del Catacumbo, Guayabo area, and Encontrados, to go from 1200–1600 to 800–1200 mm/year	• African palm, banana, cocoa, and grassland (swamps) • Livestock and forestry plantations • Coffee and vegetables (colon)	Mainly in the perennial crops, there are areas that go from medium to marginally suitable (south of the lake: musaceous, palm, and cocoa) Livestock could be affected by heat stress
Andes	The CCC-EQ model that the area from 1200–1600 to 800–1200 mm/year and this in turn extends significantly toward the foothills of Barinas and Apure and South Táchira, this increases in the drier area that goes from south of the city of Barinas to Guasdualito affecting the forests of Ticoporo, San Camilo, and El Nula	• Livestock and forestry plantations • Cocoa • High altitude farming and livestock • Corn-Leguminous • Coffee and vegetables	Mainly in the perennial crops, there are areas that go from medium to marginally suitable (Tachira's Andes: coffee)

Source Adapted from Venezuela (2005)

Andean Development Corporation/Latin American Development Bank (CAF 2014). Guatemala, El Salvador, Honduras, Belize, and Nicaragua appear with an extreme vulnerability index, while Panama shows a medium index and Costa Rica is the only country in the region with a low IVCC. This vulnerability is greater for the countries most dependent on agriculture, mainly in the case of Guatemala, which is the country with the highest risk of vulnerability in the Central American region, ranking second among the LAC indices after Haiti (CAF 2014). Both Central America and the Caribbean region also have the highest vulnerability indexes of all LAC with respect to climate-related extreme events (IPCC 2014b; CAF 2014).

The Global Climate Risk Index for 2014 also placed Honduras (1st), Nicaragua (5th) and Guatemala (10th) among the ten countries most affected by climate events between 1995 and 2014, as shown in Map 3.28 (German Watch 2016).

One of the greatest climatic risks in the region is the drought caused by various phenomena, including El Niño (ENSO), which causes serious threats and losses of agricultural production. In addition, in recent years the frequency of torrential rains and storms has increased, complicating the vulnerable situation of areas on hillsides with different levels of deterioration, which in turn cause flooding and alluviums that

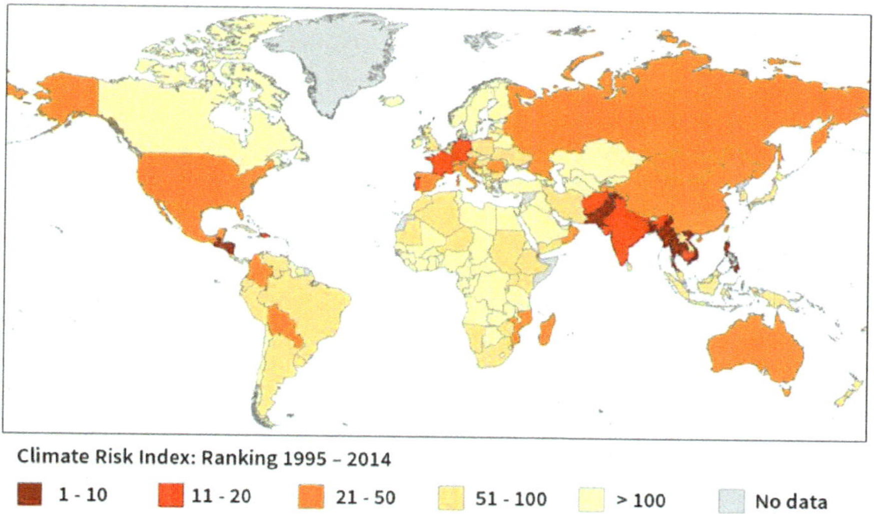

Map 3.28 Global climate risk 2014. *Source* German Watch (2016)

affect works and human activities, mainly in the lower parts (IPCC 2014b; German Watch 2016; CAF 2014). Thus, changes in the rainfall regime are one of the main drivers of risk in Central America, while it is likely that the lower incidence of rainfall in the region will occur at the same time as a higher frequency of extreme rainfall occurrence, leading to a greater number of episodes of both droughts and floods (IPCC 2014b; CAF 2014).

In view of this situation, in which drought is one of the most frequent natural hazards in the Pacific coastal areas of Central America, the World Food Programme (WFP) of the United Nations (UN) has identified the subregion of the Central American Dry Corridor (CSC).[13] This corridor is made up of dry areas (with 6 or more months of dry season) adjacent to each other found in parts of Guatemala, Honduras, El Salvador, Nicaragua, Costa Rica, and Panama, where there is a high social and environmental vulnerability to the process of climate change and climate variability (FAO 2012; Vega García 2008; CATHALAC 2010; Bouroncl et al. 2015) This corridor was conceived by combining climate data obtained in areas affected by droughts in 2000–2001 and is specifically made up of rural areas in four countries: Guatemala, El Salvador, Honduras, and Nicaragua. However, due to the characteristics of some dry areas in Costa Rica (Guanacaste, Herrera and Los Santos) and Panama (Veraguas), these can also be considered as part of the corridor. See Maps 3.29 and 3.30.

This corridor is part of the dry tropical forest eco-region that covers the lowlands of the Pacific slope and much of the central region of the piedmont (0–800 m) of El Salvador, Honduras, Guatemala, Nicaragua, and Guanacaste in Costa Rica. The

---

[13] It is called indistinctly Central American Dry Corridor, Semi-arid Belt of Central America or Central American Drought Corridor, being all of them similar denominations (in practice).

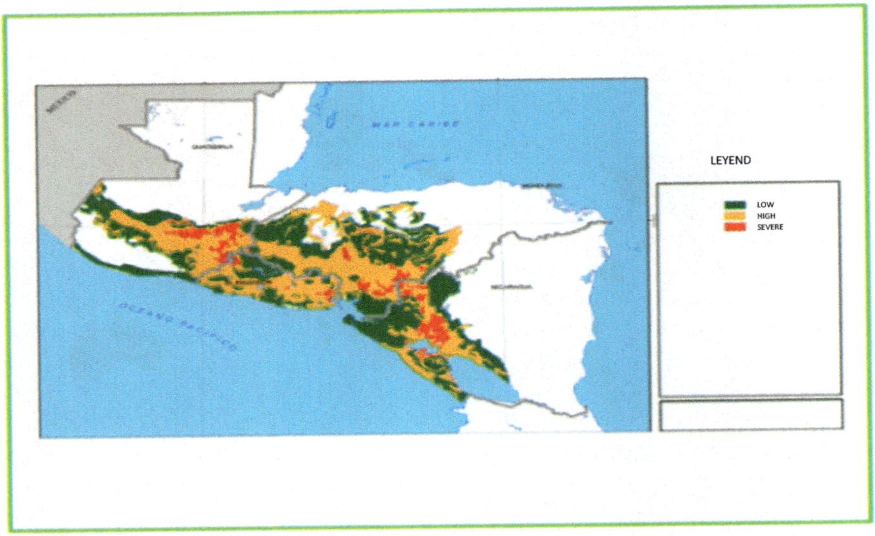

Map 3.29 Central American dry corridor. *Source* Adapted from FAO (2012). Based on the integration of dry months map—CIAT, PREVDA Atlas-isohyets, regional map and Holdridge life zones

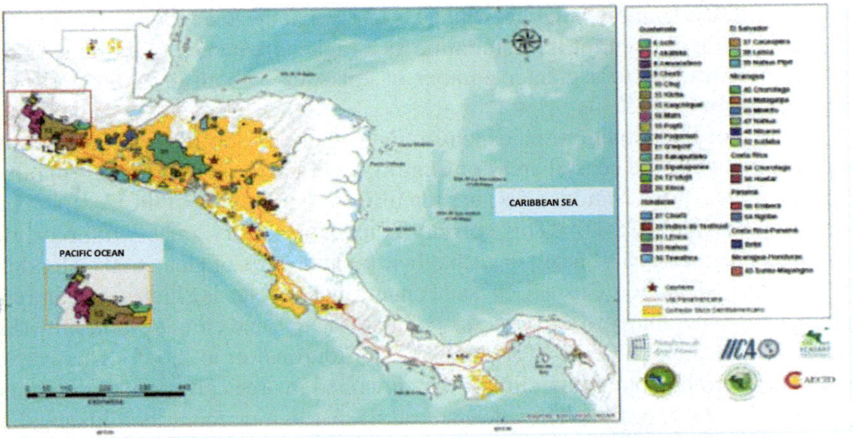

Map 3.30 Native communities established in the semi-arid belt of Central America. *Source* Adapted from CATHALAC (2010)

CSC has a marked and prolonged dry season (summer), and during the rainy season (winter) there is a latent risk of recurrent droughts that occur for different reasons: a late entry of winter, a prolongation of the heat wave or a premature suspension of winter. In this corridor the drought is cyclical and closely related to the El Niño period. In the last 60 years, about 10 El Niño events have been observed with a variable duration of 12–36 months. Periods of recurrence are unpredictable and there is no evidence yet that their frequency has increased as a result of climate change and global warming (FAO 2012; IPCC 2014b).

In Map 3.29, it can also be seen that in this area, three zones are differentiated, depending on the effect of the drought: severe effect, high effect, and low effect. Those of severe effect are the most susceptible. In those three zones converge a long dry season (more than six months), low precipitation (800–1200 mm/year and even less in specific places), and high evapotranspiration. The natural vegetation are savannahs and shrub forests. In high effect areas, the average precipitation is 1200–1800 mm/year the dry season lasts 4–6 months, and the evapotranspiration is medium. Pine forests and mixed multilevel forests are added to the vegetation. The low effect zones, although they are located in the dry corridor, have a high precipitation of 1600–2000 mm/year. The dry season lasts between 4 and 6 months, and evapotranspiration is low. In addition, there are submontane, montane, and highland evergreen forests, as well as riparian forests (FAO 2012; Vega García 2008).

Within the CSC, the two countries with the highest proportion of severe impact zones in their territory are Guatemala and Nicaragua (11.5%), while El Salvador and Honduras have 4% and 3.9%, respectively. However, the countries with the largest proportion of their territory in the dry corridor are El Salvador (with 100%) and Honduras (with more than half). In Guatemala and Nicaragua, meanwhile, the proportion is less than 50%.[14] The CSC also has the highest population density of these four countries. In this sense, the area is inhabited by more than 8.6 million people, of which 2.5 million are in Guatemala (94 municipalities, distributed in 16 departments); 1.2 million in El Salvador 1.2 (95 municipalities in 4 departments); 2.2 million in Honduras (202 municipalities in 10 departments) and 2.6 million in Nicaragua (also 94 municipalities in 16 departments).

One of the characteristics of this corridor is the frequent occurrence of natural disasters that make its inhabitants very poor and suffer from chronic malnutrition (in addition to problems of access to water) as they face at least six months of drought a year. In addition to droughts, in the last decades they have suffered other extreme events of natural causes (climatic and non-climatic) such as hurricanes, floods, earthquakes, volcano eruptions, landslides and, in some periods, also economic effects, such as the coffee crises, which aggravates the deteriorated prevailing socioeconomic situation (Vega García 2008).

The chain of these situations means that the population does not have time to recover by continually starting from scratch, generating a vulnerability that is cumulative and has migration as an "immediate" adaptation measure. According to

[14] In the case of Costa Rica and Panama, the percentages are smaller than these figures.

the WFP (World Food Programme 2002; Vega García 2008), the characteristics of vulnerability are:

- *Tendency to suffer natural disasters*: Drought is recurrent. While expected, farmers cannot handle it. Thus, families lose the ability to cope with successive events and are unable to recover from losses. Food security is precarious because each impact affects nutrition and consumption patterns.
- *Generalization of losses*: In recent years, most families have been affected (85%). Drought causes loss of crops, seeds, and animals, reduces job opportunities, and leads to low income, unemployment, and migration (either for long or short periods).
- *Poverty. Reduced physical and human capital*: Precarious conditions of education (illiteracy and incomplete education), health, and public services.
- *Dependence on subsistence agriculture*: The main economic activity (main livelihood of the inhabitants) is subsistence agricultural production, complemented by some jobs (agricultural and non-agricultural), livestock, and small businesses. Most families do not have access to land.
- *Loss of assets and altered life patterns*: Families have been losing their assets as a result of the continuous negative impacts. It is increasingly difficult for them to re-establish a minimum of viable means of subsistence. People sell their animals and migrate for a short time. In general, this is not a sustainable mechanism. Changes are made in the diet (number of meals, children's diet, changes in products, and reduction of quantities).
- *Vulnerable groups in the midst of subsistence*: Landless families are the most sensitive group. Within families, children, and single parents. Reinforcement of poverty by continuous depletion of family capital and the difficulty in restoring a viable livelihood level, which prevents families from responding effectively to changes.

In this way, migration, which in the region has traditionally been due to socioeconomic and political reasons, is presenting an environmental component that lately is becoming increasingly relevant (Vega García 2008). Historically, Central America has been affected by a series of environmental events that highlight its high vulnerability. Such is the case of El Niño of 1997–1998 (when Hurricane Mitch occurred), or in 2001 when there were large agricultural losses as well as the closure of several coffee farms, which increased food insecurity and triggered famines in some extreme cases, such as the case of Matagalpa in Nicaragua. Such episodes of food crises led to population displacement within the various Central American countries. This shows that migration in the region is also influenced by environmental factors and can be amplified by climate change, which generates new challenges in terms of governance. Therefore, the identification of vulnerable regional spaces becomes a necessity in the face of potential environmental migration scenarios, beyond the borders of the respective countries. In this sense, it is necessary to develop greater knowledge and elaborate a theoretical-conceptual framework to include this environmental migration in regional public policies and in legal frameworks that can regulate it (Vega García 2008). Map 3.30 shows the diversity of communities that are involved in this problem, which makes the analysis and management of these new situations even more complex.

An additional factor to consider is that, in Central America, these drought-prone areas coincide mainly with areas of basic grain production (corn, sorghum, beans, and rice) as well as with areas where important forestry and livestock activities take place, and for the cultivation of seasonal products that are relevant to the region's exports, such as coffee. In this sense, the coffee sector and workers in the sector face a double threat due to the exposure to the economic collapse induced by international market prices in addition to climatic factors (CATHALAC 2010; Vega García 2008; FAO 2012; ECLAC 2010).

The importance of the SSC from various points of view led to the corridor being studied by various organizations operating at the regional level,[15] including ECLAC and FAO.

According to ECLAC (2010), agriculture continues to be a very important activity in Central America, especially in Nicaragua, El Salvador, Guatemala, and Honduras, occupying large territorial areas: 70% in Costa Rica, 68% in El Salvador, 53% in Honduras, 50% in Guatemala, 47% in Nicaragua, 38% in Panama, and 17% in Belize. Approximately 130,000 ha are under irrigation (7.3% of the agricultural area), with large differences between countries, Costa Rica being the country with the largest irrigated area. The agricultural sector is one of the main engines of the region's economy, accounting for approximately 11% of total GDP. Considering the contribution of agribusiness, this percentage rises to 18%. The sector is also the main supplier of food and inputs for industry, contributing 35% of total exports. This sector (and the rural milieu in general) absorbs a significant part of the economically active population, representing an important source of income for rural households. However, production is slow, while yields have remained stagnant, affecting its competitiveness and growth potential. The low productivity is explained by the low capitalization and the damage caused by climatic phenomena, among other factors (ECLAC 2010).

It emerges from FAO (2012) that in the CSC, corn and bean producers produce on average 1.3 ha in El Salvador and Guatemala; 2.4 ha in Honduras, and 2.8 ha in Nicaragua. These are small producers who produce for family consumption and market part of their production to cover their non-food needs. The income of these small producers is around USD 72 per month in Honduras and up to USD 104 per month in El Salvador. These incomes are complemented by other sources, sometimes including other family members, to round out an average of USD 177 per month, thus saving the line of extreme poverty or indigence but still remaining below the poverty line. In this subsistence agriculture, in general, income is a function of the sale of their labor as agricultural day laborers. Both situations contribute to feeding the highest levels of vulnerability because the same factor that affects their own production tends to affect their source of employment (FAO 2012).

[15] The Regional Committee on Water Resources (CRRH), the United Nations Economic Commission for Latin America and the Caribbean (ECLAC), the Coordination Center for the Prevention of Natural Disasters in Central America (CEPREDENAC), among others.

There is a very particular situation regarding the vulnerability of the SCC. The region has sufficient natural resources to ensure the food and nutrition of more than 30 million people; however, it has some of the worst levels of inequality in Latin America and the world. These countries have a per capita GDP ranging from USD 2000 to USD 4000, the primary sector has a share of up to 20% of GDP, and the levels of poverty and malnutrition are alarming and affect mainly rural populations (see Table 3.4), where children are especially vulnerable (FAO 2012). In Guatemala almost half of the children under 5 years old suffer from moderate to severe chronic malnutrition, while these percentages are 29% in Honduras; 22% in Nicaragua 22%, and 19% in El Salvador 19%. This is aggravated by the environmental degradation suffered by the region, mainly through deforestation that causes erosion and soil degradation, increased frequency and impact of floods, contamination and sedimentation of water bodies, drying of rivers during part of the year and landslides (FAO 2012). These four countries are the ones that show the most notable decrease in forest coverage as a percentage of the national territory between 2005 and 2010 (2.1% in Nicaragua, 2% in Honduras, 1.5% in Guatemala, and 1.5% in El Salvador) (FAO 2012).

In the SCC, the greatest vulnerability to drought is experienced by small-scale subsistence agriculture and small rural communities due to reduced sources of drinking water. This makes drought a major socioeconomic phenomenon because of its effects on the livelihoods of the rural poor. Among the factors that accentuate vulnerability is the degradation of soils with the lower water retention capacity that it entails, as well as loss of fertility, loss of infiltration capacity, and crop yields and greater susceptibility to the impact of heat waves. In Guatemala, the degradation of 12% of the territory has had an economic cost that exceeds US $2 billion and mainly affects 100,000 families in conditions of extreme poverty (FAO 2012).

Map 3.31 shows the modeling results for the region at the end of the twenty-first century (ECLAC 2010). In the base scenario (year 2005), 41% of the region's land was dedicated to agricultural use, 43% corresponded to forest, 12% to savannahs, shrubs and natural grasslands, 0.5% to urban use, and almost 4% to other uses. According to this modeling, by 2100 one could expect the loss of approximately one third of 2005s forests and up to 80% of grasslands, savannas, and shrubs, while the agricultural area would grow to 50% (see second image of Map 3.31). Most of these changes would occur by 2050. Since the currently most fragmented forest areas have historically been the most affected (i.e., the most threatened), the model estimates that

Table 3.4 Poverty and indigence (urban and rural) in the CSC

Country	Rural poverty	Urban poverty	Rural indigence	Urban indigence
Honduras	78.8	56.9	61.7	26.2
Nicaragua	71.5	54.4	46.1	20.8
Guatemala	66.5	42.0	42.2	14.8
El Salvador	57.6	42.3	25.2	12.8

Source Adapted from FAO (2012)

these forest areas will be deforested first. It is important to note that this is a Business as Usual scenario, that is, one in which no specific actions are proposed to change the currently predominant trends, but which also does not consider the potential impacts of climate change. It must be taken into account that climate directly influences the growth and development of plants and crops, water balances, and land erosion. The El Niño phenomenon on the Pacific side of Central America has decreased rainfall levels, delayed the rainy months, and increased the average temperature as well as reduced cloudiness, causing longer summers between July and August and more sunshine (ECLAC 2010).

Maize cultivation in Central America can therefore be expected to suffer significant decreases, in some cases up to 15% of current production. A general reduction in rice productivity is also expected and may reach up to 31% in Costa Rica. For Guatemala, a scenario with a temperature rise of 3.5 °C and a 30% drop in rainfall resulted in decreases of up to 34% for maize, up to 66% for beans, and up to 27% for rice. In Costa Rica, it was determined that rice, potato, and bean yields will decrease, while coffee yields will increase as temperatures rise. In Panama, corn yields are expected to decrease by 34% in 2050 and fall to 21% in 2100, all with respect to current productivity. For Honduras, a study determined that corn yields will be reduced by 22% by 2070 (ECLAC 2010).

Table 3.5 shows some results of the expected economic impacts of climate change on agriculture in Central America for different scenarios (A2 and B2) and for different discount rates, showing the strongly negative impacts expected on the agricultural sectors and, consequently, on the economies of the Central American countries, given the importance of the sector in their economies (López Feldman 2016).

Corn, beans, and rice are fundamental in providing calories and protein to large parts of the Central American population. Depending on the country and the grain, there is significant self-consumption production by poor small farmers. The effect of climate change on agriculture will have a significant impact on food security and poverty, by reducing food production and direct access to it by rural producers. If there were a price increase and/or a situation of shortage, the situation would be even more complicated, depending on the possibilities of compensatory imports (ECLAC 2010).

Clearly, the results highlight the need for immediate action by taking adaptation measures at local, national, and regional levels. At these levels, the agricultural response to climate change will require close coordination with policies to reduce deforestation, protect biodiversity and manage water resources. Thought will have to be given to expanding experiences with more sustainable production processes, such as agroforestry, the promotion of native varieties, the combination of agricultural activities with those that protect natural ecosystems, and eventually payment for environmental services, in an effort to strengthen the well-being of rural and indigenous populations (ECLAC 2010).

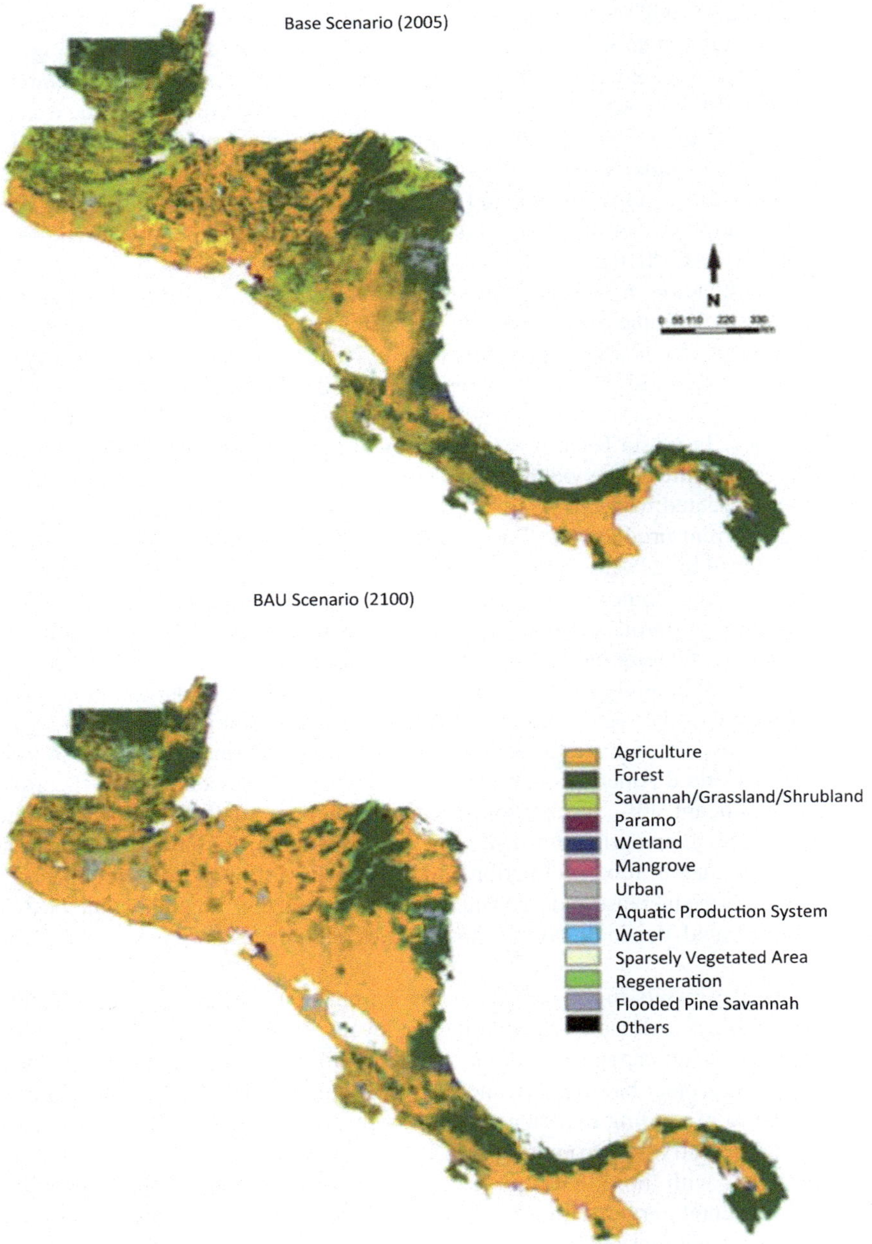

Map 3.31 Land use change scenarios in Central America (2005–2100). *Source* Adapted from ECLAC (2010)

Table 3.5 Climate change effects on agriculture according to different scenarios in the CSC

Effects of climate change on agriculture in different scenarios						
Author	Country	Economic effect of changes in precipitation and temperature (2020–2100) and in agricultural production (percentage of 2007 GDP)				
		2020	2030	2050	2070	2100
Mora et al. (2010b)	Guatemala	(– 0.03, 0.45)	(0.02, 0.78)	(– 1.30, 1.01)	(– 4.83, 0.37)	(– 18.77, – 0.45)
Ordaz et al. (2010b)	Costa Rica	(– 1.22, – 0.14)	(– 2.08, – 0.44)	(– 3.51, – 1.46)	(– 5.87, – 2.56)	(– 11.86, – 3.67)
Ordaz et al. (2010a)	El Salvador	(– 1.08, – 0.91)	(– 2.06, – 1.29)	(– 2.62, – 1.66)	(– 3.22, – 1.91)	(– 7.99, – 2.27)
Mora et al. (2010a)	Panama	(– 4.01, 1.48)	(– 5.52, – 2.46)	(– 10.04, – 4.26)	(– 13.25, – 5.86)	(– 19.42, – 7.61)
Ramírez, Ordaz, Mora, Acosta and Serna	Nicaragua	(– 3.37, – 0.38)	(– 5.52, – 2.04)	(– 8.58, – 2.25)	(– 11.66, – 2.63)	(– 22.48, – 2.96)
Ordaz, Ramírez, Mora, Acosta and Serna (2010)	Honduras	(– 2.30, – 1.02)	(– 4.00, – 2.08)	(– 7.69, – 2.76)	(– 10.94, 3.30)	(– 18.60, – 3.87)
Ramírez, Ordaz, Mora, Acosta and Serna	Belize	(– 6.42, – 0.94)	(– 12.25, – 3.84)	(– 16.78, – 4.93)	(– 21.40, – 6.09)	(– 34.50, – 7.20)

Source Adapted from López Feldman (2016)

75% of the population of Central America is supplied with groundwater. The largest aquifers are those of Nicaragua (Managua) and Costa Rica (north of the Central Valley), where half of the population uses them as a source of drinking water. Many aquifers on the Pacific side of Honduras, El Salvador, and Guatemala have high salinity, which has increased significantly since 2005. Their future use is threatened by the possibility of rising sea levels. In the aquifers of San Salvador (El Salvador), Managua (Nicaragua), and San José (Costa Rica), water quality problems have been identified due to the infiltration of agrochemicals and untreated domestic wastewater (ECLAC 2010).

With respect to water resource uses, there is a wide range of availability per capita, between El Salvador with barely 1752 m^3 per inhabitant per year, a level close to that of water stress, and Belize with 66,429 m^3 (ECLAC 2010). Total regional extraction is 12.2 billion m^3 per year; Guatemala consumes 42% of this total, followed by Costa

Rica with 22%. In Honduras, Guatemala, Costa Rica, and El Salvador, between 83 and 54% of their reported national extraction is devoted to agriculture, while in Panama industrial consumption dominates, with 66%. Despite the high levels of water availability, the population in many areas of Central America suffers from shortages. The seasonal imbalance between water availability and demand has caused river runoff to be limited to the rainy season in some areas, leaving rural areas without water sources for half of the year. Pollution also limits water availability in urban and rural areas and increases the cost of supply (ECLAC 2010).

The worrying fact is that future prospects for resource availability show that it remains at current levels until about 2030 and reductions are significantly higher in the last three decades of the century, especially in climate scenarios where further temperature increases are expected. In these scenarios, by 2100 the availability in the region is reduced by 63% with respect to the year 2000 (35% if scenarios with lower expected temperature increases are taken). El Salvador is the country with the greatest loss by 2100 (between 50 and 82%, depending on the scenario). The country with the smallest reductions is Panama, but even so they would be 13% and 51, respectively. For all countries, around 2025 the total renewable availability begins to reduce the ecological volume. See Graph 3.1 and Map 3.32 and (ECLAC 2010; WRI 2016).

Within the complex panorama presented by the region, perhaps the most critical situation is that of Guatemala, mainly in terms of "seasonal hunger", a process defined as a predictable and recurrent seasonal worsening of the food and nutritional situation of vulnerable households, related to climatic and health factors, and to agricultural cycles. The most relevant indicator to measure the evolution of seasonal hunger

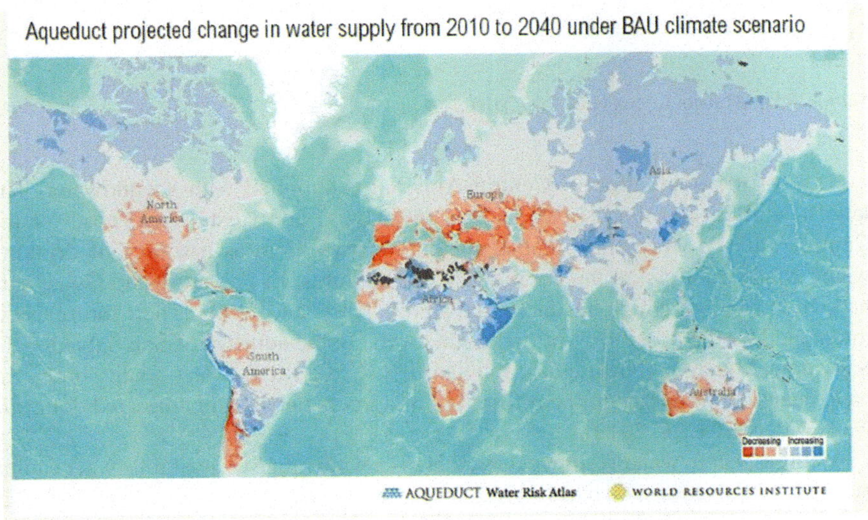

Map 3.32 Changes in the availability of water 2010–2040. BAU scenario. *Source* WRI (2016)

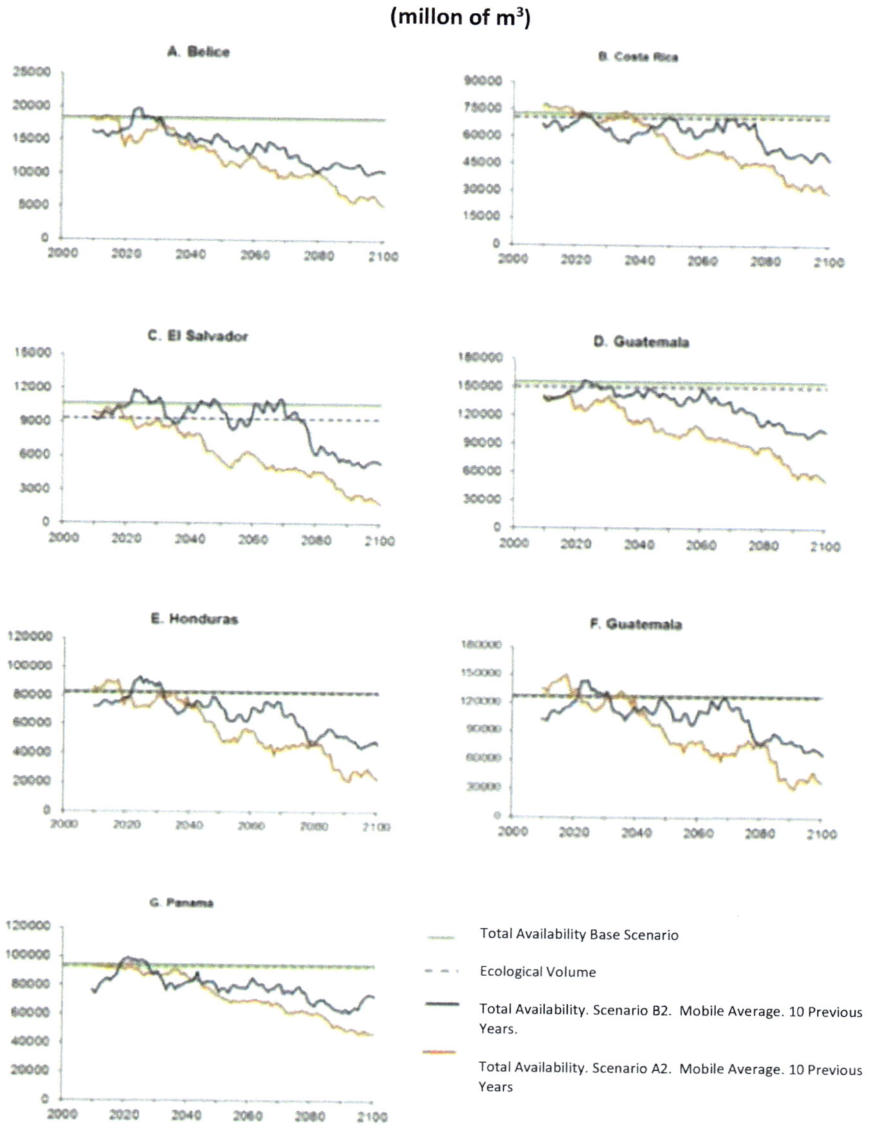

Graph 3.1 Central America: evolution of renewable total availability of water according to base, A2 and B2 scenarios; 2000–2004 to 2100. *Source* Adapted from ECLAC (2010)

in Guatemala, which affects thousands of households in the country, is child acute malnutrition (Vivero Pol 2013). Seasonal hunger manifests itself in various ways: (a) reduced meal times, (b) impoverished food diversity, (c) reduced food reserves at the household level, or (d) increased dependence on food purchased outside the home. On the other hand, the variables affecting seasonal hunger are as follows: (1) the rainfall regime, (2) extreme weather events, (3) global acute malnutrition, (4) diarrheal diseases, (5) respiratory infections, (6) the agricultural calendar for basic grains, (7) national market prices of basic grains, (8) household basic grain reserves, and (9) temporary work in agricultural tasks (Vivero Pol 2013).

There are two marked periods of seasonality of hunger and its main indicators: a period from January to August, when acute malnutrition is increasing and the seasonal factors that trigger it are activated; and a period from September to December, of recovery and return to minimum values. In this sense, climate change will affect hunger. Seasonal hunger and the accompanying peaks in child acute malnutrition are regular and predictable and therefore preventable and mitigable. Extreme weather events (droughts and hurricanes) are also seasonal but with a more irregular and spaced frequency. These phenomena will increase their frequency and severity as a consequence of climate change and global warming and will strongly affect the livelihoods of vulnerable households. There will be more and more torrential rains and hurricanes affecting agricultural production in Guatemala and droughts will be more intense. In addition, the seasonality of the rainfall regime will also be affected, which may indicate an increase in the relevance of seasonal hunger for the coming decades (Vivero Pol 2013).

In addition, the projected increase in average annual temperature and decrease in precipitation will also have other significant impacts on Guatemalan agriculture, as the areas suitable for crops that support agricultural exports and peasant food security are likely to change in the future. Some municipalities will gain productivity for certain crops while others will lose it. The capacity of rural people to adapt to these changes (whether they represent a loss or a gain) will depend on their access to basic services, information, resources needed for innovation, and the ability to maintain healthy ecosystems (Bouroncl et al. 2015). It is expected that the increase in temperature accompanied by the intensification of dry and hot periods and less rainfall will cause a water deficit and consequently a change in areas suitable for cultivation. Of the crops analyzed, the most sensitive to changes in climate are beans and coffee, with a decrease in the area of cultivation of both being expected throughout the country. In the case of coffee, decreases are foreseen in all the municipalities where it is now grown. More suitable areas are appearing (fundamentally those at higher altitudes), while the lower areas (Atlantic coast, Pacific coastal plain and dry corridor) are losing their suitability and the same is happening with beans, sugar cane, and corn. One of the problems is that crop gains in the highlands will occur where urban land use or conservation for the provision of ecosystem services (such as water) would conflict with potential expansion of agriculture (Bouroncl et al. 2015). The vulnerability to climate change of the agricultural sector varies from municipality to municipality depending on sensitivity, exposure and adaptive capacity. The more the rural population depends on economic income from agriculture, the more sensitive it will be to the effects of climate change on crops. On average, 70% of the rural population over the age of 15 has agriculture as its main job, with higher percentages in the country's most remote municipalities. The agricultural sector has a very high proportion of labor that is very vulnerable to changes in crop productivity and production because they are poor small- and medium-sized producers in a situation of social inequality. This situation will be exacerbated by climate change. It is estimated that in 2030, 8% of the areas suitable for their current crops will be lost. Most of this workforce has a low level of education. To increase the adaptive capacity, these human resources and local knowledge management must be strengthened (Bouroncl et al. 2015).

3.9 Conclusions and Recommendations

Although the four areas analyzed have differences between them, there are issues that can be considered as common to all of them.

First, for many countries in Central and South America a priority measure for adaptation to future climate change is to reduce vulnerability to the current climate (IPCC 2014a; ECLAC 2010, 2014a).

In this sense, there are examples that show synergies between socioeconomic development, the adaptation and mitigation planning, which can help governments and local communities to efficiently allocate available resources in the design of strategies to reduce vulnerability. However, mainstreaming such actions on a continental scale requires that development policies and climate stress adaptation strategies are truly intertwined in order to achieve vulnerability reduction. Enhancing institutional and technical capacities and economic resilience (mainly to price changes and market volatility) is crucial for successful adaptation to climate change. It is necessary to increase the capacity and skills to adjust to the changes proposed, not only by the climate but also by a changing economic and geopolitical context.

With these forecasts, there is a strong demand for research to evaluate the effects of climate change in the different areas with semi-arid climates of Central and South America. Research is a key factor in finding resilient climate adaptation pathways and in generating knowledge to apply technologies, processes, and actions to reduce or manage climate risks in the region. In this sense, it is of crucial importance to raise awareness about the need for a comprehensive vision of the problem of adaptation (Angelotti et al. 2015). Planning, both at regional, national, subnational, and local levels, is important, but fundamentally the strengthening of the capacities of local actors and organizations that, together with the extension services of national entities, are those that will transfer to the territory the strategies, policies and measures that are expected to provide as a result a greater resilience to current conditions and changes.

With regard to agriculture (ECLAC 2010, 2014a; IPCC 2014a), it is noted that global warming can extend growing seasons, so changing planting dates is an option frequently identified for cereals and oilseeds as long as there is no increase in drought at the end of the growing season. Higher temperatures reduce yields and quality so improving tolerance to high temperatures is often an adaptation option identified for almost all crops. Improving genetic conservation and expanding access to gene banks can also facilitate the development of better adapted crop varieties, as well as the development and cultivation of varieties resistant to water-stressed conditions in the face of increased drought. Adaptive techniques in water management include improved storage and access to irrigation and more efficient water supply systems, improvements in irrigation technologies such as drip irrigation, improved efficiency in water harvesting and collection, agronomic practices that increase water and moisture retention in the soil through practices such as minimum tillage and tree canopy management.

Specific adaptation measures being considered include, in summary, the following (see, among others: IPCC 2014a; Angelotti et al. 2015; REGATTA 2013):

- Adaptation actions supported by the integration of various biological diversification measures to combat desertification.
- Ecosystem-based adaptation actions (establishment of protected areas), as well as community-based conservation and management of natural areas.
- Adaptation actions in agriculture, through rainwater harvesting, efficient use of irrigation water, conservation of soil moisture through the use of dead plant cover, genetic improvement with selection of materials resistant to drought and high temperatures, application of multiple crop systems (or polycultures), use of local genetic diversity, among other measures.
- Adaptation actions related to the water issue, through the application of management techniques to increase water use efficiency: the use of subway storage, emergency irrigation, and drip irrigation.
- Adaptation actions in the agricultural techniques used to minimize the impacts of agriculture, such as organic fertilization, as well as the use of stubble both to increase the capacity to retain soil moisture and to reduce evapotranspiration losses. The use of spontaneous vegetation is also recommended as an alternative to dead cover (scarce, in general in semi-arid areas) that can increase water use efficiency.
- In the case of monoculture, temporary measures can be applied such as altering sowing dates, introducing new cultivars, and using irrigation.
- An alternative is the diversification of crops, seeking that the heterogeneity in the scale of the landscape can effectively help to increase agricultural production.
- The development of silvopastoral and agroforestry systems. These systems contemplate the integration of diverse livestock species with commercial systems of permanent forest species and also in dry farming systems. As an additional effect, it is verified that the agroforestry system reduces the loss of soil, water and nutrients and, consequently, delays the erosion process; being also useful to accumulate carbon. These agrosilvopastoral systems can consist of areas with introduction or maintenance of the arboreal component (native or exotic) with cultivated pastures adapted to the semi-arid and systems with use of corn, grasses, and fodder legumes. Or these agrosilvopastoral systems can also be developed through the integration of tree species (native or exotic) with crops adapted to the semi-arid such as cassava or sorghum.

- Opportunities for the development of these areas can be explored by applying solutions linked to payment for environmental services.
- Public policies of an environmental nature are not the exclusive preserve of the State. All actors in society (public, private, non-governmental organizations, universities, and the population in general) must participate in order to achieve legitimacy.
- Environmental issues are long-lasting, often exceeding government cycles.
- Governments must cooperate in the transfer and use of technologies that allow the problems that make climate variability, are collected with the least possible margin of error.
- Decisions that are not sustainable in a context of climate variability and that generate poor adaptation and lead to pejorative situations in the long run should be avoided.
- As seen previously in other parts of this document, a particular case is that of the Semi-Arid Corridor in Central America (see FAO 2012; Vivero Pol 2013; among others). The main objective is to generate prevention mechanisms for the most vulnerable households, in a context where extreme weather events increase the rates of acute and chronic child malnutrition.

The Regional Strategic Framework for Climate Risk Management in the Central American Semi-Arid Corridor (MERGERCA) is an initiative of the project to increase the resilience of small-scale producers' livelihoods to drought in the CSC (FAO 2012). The main objective of the project is to provide a strategic framework to promote coherent, comprehensive, and participatory actions to reduce vulnerability in risk management and increase the resilience of agriculture and natural resources (water, soil, forest, biodiversity, and landscape) and contribute to food and nutritional security in the face of extreme climate events. It seeks to implement successful practices to develop capacities, sustainability, employment, income, food security, and a better quality of life. Six axes: (1) strengthening of capabilities, (2) land-use planning and integrated watershed management for the management and prevention of climate risks, (3) research and transfer of technology and agricultural practices, (4) scaling up of successful experiences and practices in climate risk management (CRM) in agriculture, (5) harmonization of regional, national and local policies, strategies, plans and programs linked to CRM, and (6) identification and management of innovative national and multilateral financial sources for CRM.

Acronyms

4AR	IPCC Fourth Assessment Report (2007)
5AR	IPCC Fifth Assessment Report (2013–2014)
ACF	International Foundation Action Against Hunger
BSR	Business for Social Responsibility

CAF	Andean Development Corporation. Latin American Development Bank
CAPECO	Paraguayan Chamber of Exporters and Marketers of Oilseeds
CATHALAC	Water Center for the Humid Tropics of Latin America and the Caribbean
CATIE	Tropical Agricultural Research and Higher Education Center
CCAFS	Research Programme on Climate Change, Agriculture and Food Security
CCDA	Central American Commission on Environment and Development
CEPREDENAC	Coordination Center for the Prevention of Natural Disasters in Latin America
CGIAR	Consultative Group on International Agricultural Research
CI	Conservation International (International Conservancy)
CIAT	International Center for Tropical Agriculture
CISL	Institute for Sustainable Leadership
CONRED	National Coordination for the Reduction of Natural or Provoked Disasters
COPECO	Permanent Contingency Commission (Honduras)
CRRH	Regional Committee on Water Resources of the Central American Isthmus
CSC	Central American Dry Corridor
ECF	European Climate Foundation
ECHO	Directorate-General for Humanitarian Aid and Civil Protection of the European Commission
ECLAC	United Nations Economic Commission for Latin America and the Caribbean
ENSO	El Niño-Southern Oscillation
EU	European Union
FAO	Food and Agriculture Organization of the United Nations
GEF	Global Environmental Facility
IARNA	Institute of Agriculture, Natural Resources and Environment (Universidad Rafael Landívar-Guatemala)
IDB	Inter-American Development Bank
IGEC	Polluting Power Generation Index
IHP-LAC	International Hydrological Programme—Regional Bureau for Science in Latin America and the Caribbean
IICA	Inter-American Institute for Cooperation on Agriculture
INCI	Interciencia Magazine
IPCC	Intergovernmental Panel on Climate Change
IVCC	Climate Change Vulnerability Index
LAC	Latin America and the Caribbean
MARN	Ministry of Environment and Natural Resources (Bolivarian Republic of Venezuela)
MDGs	Millennium Development Goals Achievement Fund

MMA	Ministry of the Environment (Federative Republic of Brazil)
MMARNyDF	Ministry of Environment, Natural Resources and Forest Development (Bolivia)
MMAyA	Ministry of Environment and Water (Plurinational State of Bolivia)
NAMAs	Nationally Appropriate Mitigation Actions
NCCP	National Climate Change Program (Plurinational State of Bolivia)
NCDs	Intended Nationally Determined Contributions
NY	New York
PREVDA	Regional Programme for the Reduction of Vulnerability and Environmental Degradation
RCP	Representative Concentration Pathways
REGATTA	Regional Portal for Technology Transfer and Action on Climate Change in Latin America and the Caribbean
SAyDS	Secretariat of Environment and Sustainable Development of the Nation (Argentine Republic)
SEAM	Secretariat of the Environment (Republic of Paraguay)
SICA	Central American Integration System
SINAGER	National Risk Prevention System (Honduras)
SRES IPCC	Special Report on Emission Scenarios
SREX IPCC	Special Report on Managing the Risks of Extreme Events and Disasters to Advance Climate Change Adaptation
SRREN IPCC	Special Report on Renewable Energy Sources and Climate Change Mitigation
UCJBS	University of Cambridge's Judge Business School
UKaid	British Department for International Development (DFID)
UN	United Nations
UNDP	United Nations Development Programme
UNEP	United Nations Environmental Programme
UNESCO	United Nations Educational, Scientific and Cultural Organization
UNFCCC	United Nations Framework Convention on Climate Change
USA	United States of America
WB	World Bank
WFP	Word Food Program (see WFP)
WG I	IPCC Working Group I (The Physics of Climate Change)
WG II	IPCC Working Group II (Impacts, Vulnerability and Adaptation)
WGIII	IPCC Working Group III (Climate Change Mitigation)
WFP	World Food Programme of the United Nations
WRI	World Resource Institute

References

Adams R, Hurd B, Lenhart S, Levy N (1998) Effects of global climate change on agriculture: and interpretative review. Clim Res 11(1):19–30

Angelotti F, Signor D, Giongo V (2015) Climate change in the Brazilian semiarid: experiences and opportunities for development. Braz J Phys Geogr 8(Special Number IV SMUD):484–495. www.ufpe.br/rbgfe

Birthal PS, Khan MT, Negi DS, Agarwal S (2014) Impact of climate change on yields of major food crops in India: implications for food security. Agri Eco Res Rev 27(2):145–155

Bouroncl C, Imbach P, Laderach P, Rodriguez B, Medellin C, Fung E, Martinez-Rodriguez MR, Donati C (2015) Guatemalan agriculture and climate change: where are the priorities for adaptation? CATIE, CIAT, CI, REGATTA. CGIAR, CCAFS, Copenhagen, Denmark. http://hdl.handle.net/10568/45942

CAF (2014) Index of vulnerability and adaptation to climate change in the Latin American and Caribbean region. Caracas, Venezuela

CAPECO (2016) Sowing, production and yield area. http://www.tera.com.py/capeco/index.php?id=area-de-siembra-produccion-y-rendimiento

CATHALAC (2010) Central American atlas for sustainable territorial management. CATHALAC, UE, CEPREDENAC, IICA, CCDA-SICA, CRRH, PREVDA, Panama. Available at http://www.cathalac.int/atlas_prevda/

Díaz M (2001) Experimental ecology and ecophysiology: basis for the sustainable use of natural resources in neotropical arid zones. INCI 26(10). ISSN 0378-1844

ECLAC (2007) Climate alterations in Bolivia: impacts observed in the first quarter of 2007. ECLAC, Santiago de Chile

ECLAC (2010) The economics of climate change in Central America. In: Synthesis 2010. ECLAC, CCAD, UKAid, SICA, ECLAC Office in Mexico

ECLAC (2014a) The economics of climate change in Argentina. Synthesis report. ECLAC, Santiago de Chile

ECLAC (2014b) The economics of climate change in Paraguay. ECLAC, Government of the Republic of Paraguay (SEAM; Ministry of Treasury), Santiago de Chile

FAO (2012) Regional strategic framework for climate risk management in the agricultural sector of the Central American dry corridor. FAO, ACF, THREW OUT, Central American Dry Corridor Regional Project, Tegucigalpa, Honduras

Fischer G, Shah M, Tubiello F, Van Velhuizen H (2005) Socio-economic and climate change impacts on agriculture: an integrated assessment, 1990–2080. Philos Trans R Soc B Biol Sci 360(1463):2067–2083

Gabaldón AJ (2008) Climate change and its possible effects on Venezuela. Simon Bolivar University. In Humania del Sur. J Latin Am Afr Asian Stud 3(4):13–32. ISSN 1856-7959. University of Los Andes, Mérida, Venezuela

German Watch (2016) Global climate risk index 1995–2014. https://germanwatch.org/en/cri

Government of Argentina Republic (2007) Second national communication of the Argentine Republic to the United Nations Framework Convention on Climate Change. Head of the Cabinet of Ministers, Presidency of the Nation, SAyDS, Bariloche Foundation, Buenos Aires

Government of Argentina Republic (2015) Third national communication of the Argentine Republic to the United Nations Framework Convention on Climate Change. Head of the Cabinet of Ministers, Presidency of the Nation, SAyDS, Buenos Aires

Government of the Bolivarian Republic of Venezuela (2005) First national communication on climate change of Venezuela. MARN, UNDP, GEF/GEF, Caracas

Government of the Federative Republic of Brazil (2004) Initial communication of Brazil to the United Nations Framework Convention on Climate Change. Ministry of Science and Technology, Secretariat of Research and Development Policies and Programs, General Coordination of Global Climate Changes, Brasilia, 277 pp

Government of the Federative Republic of Brazil (2007) Atlas of areas susceptible to desertification of Brazil. MMA, Secretary of Water Resources

Government of the Federative Republic of Brazil (2010) Second national communication of Brazil to the United Framework Convention on Climate Change. Ministry of Science and Technology, Secretariat of Research and Development Policies and Programs, General Coordination of Global Climate Changes, Brasilia, 2248 pp

Government of the Federative Republic of Brazil (2016) Third national communication of Brazil to the United Nations Framework Convention on Climate Change, vol 1. Ministry of Science, Technology and Innovation, Secretariat of Policies and Programs of Research and Development, General Coordination of Global Climate Change, Brasilia, 144 pp

Government of the Republic of Bolivian (2001) National communication from the Government of the Republic of Bolivia to the United Nations Framework Convention on Climate Change. Ministry of Environment, Natural Resources and Forestry Development, National Climate Change Program, La Paz

Government of the Republic of Paraguay (2011) Second national communication on climate change of Paraguay. Presidency of the Republic of Paraguay, SEAM, UNDP, GEF, Assumption, Nov 2011

Hertel TW, Rosch SD (2010) Climate change, agriculture, and poverty. App Eco Pers Pol 32(3):355–385

IPCC (2007) Climate change 2007: impacts, adaptation and vulnerability. In: Contribution of Working Group II to the fourth assessment report of the Intergovernmental Panel on Climate Change. Cambridge University Press, Cambridge, New York, NY

IPCC (2013) Climate change 2013: the physical science basis. In: Stocker TF, Qin D, Plattner G-K, Tignor M, Allen S, Boschung J, Nauels A, Xia Y, Bex V, Midgley P (eds) Contribution of Working Group I to the fifth assessment report of the Intergovernmental Panel on Climate Change. Cambridge University Press, Cambridge, New York, NY, 1535 pp

IPCC (2014a) Climate change 2014. Impacts, adaptation and vulnerability. Summary for policymakers. In: Working Group II contribution to the fifth assessment report of the Intergovernmental Panel on Climate Change. Geneva, 32 pp

IPCC (2014b) Climate change 2014. Impacts, adaptation and vulnerability. In: Working Group II contribution to the fifth assessment report of the Intergovernmental Panel on Climate Change. Geneva, 1082 pp

IPCC (2014c) Central and South America. Executive summary. Impacts, adaptation and vulnerability, chap 27. In: Working Group II contribution to the fifth assessment report of the Intergovernmental Panel on Climate Change. Geneva, 6 pp

IPCC (2015) IPCC expert meeting on climate change, food and agriculture. In: Mastandrea MD et al (ed) Meeting report, Dublin, Ireland, 27–29 May 2015. Geneva, 68 pp

López Feldman A (2016) Climate change and agricultural activities in Latin America. In: Studies of climate change in Latin America. ECLAC, Euroclima, Buenos Aires

Martelo MT (2004) General environmental consequences of climate change on Venezuela. Caracas

McCarl BA (2010) Analysis of climate change implications for agriculture and forestry: an interdisciplinary effort. Clim Change 100(1):119–124

Mendelsohn R (2009) The impact of climate change on agriculture in developing countries. J Nat Res Pol Res 1(1):5–19

Mora J, Ordaz JL, Acosta A, Serna B, Ramírez D (2010a) Panama. Effects of climate change on agriculture. ECLAC, México, D.F.

Mora J, Ramírez D, Ordaz JL, Acosta A, Serna B (2010b) Guatemala. Effects of climate change on agriculture. ECLAC, México, D.F.

Ordaz J, Ramírez D, Mora J, Acosta A, Serna B (2010a) El Salvador. Effects of climate change on agriculture. ECLAC, México, D.F.

Ordaz J, Ramírez D, Mora J, Acosta A, Serna B (2010b) Costa Rica. Effects of climate change on agriculture. ECLAC, México, D.F.

Pluri-National State of Bolivia (2009) Second national communication of the pluri-national state of Bolivia to the United Nations Framework Convention on Climate Change. MMAyA, Vice Ministry of Environment, Biodiversity and Climate Change, PNCC, La Paz

Puche M, Silva O, Warnok R, García V (2008) Evaluation of the effect of climate change on annual crops in Venezuela. Faculty of Agronomy, Central University of Venezuela, Ministry of Environment and Natural Resources, Caracas, Venezuela. http://ceer.isa.utl.pt/cyted/venezuela2008/M_Puche.pdf

REGATTA, UNEP (2013) Study of vulnerability and impact of climate change in the Gran Chaco Americano. Progress report, June 2013. National University of Formosa (Argentina), Universidad de la Cordillera—Fundación la Cordillera (Bolivia) & Institute for Development, Participation and Citizenship (Paraguay), United Nations Environment Program (UNEP), Regional Portal for Technology Transfer and Action Against Climate Change in Latin America and the Caribbean (REGATTA), Asunción, Paraguay

REGATTA, UNEP (2014) Atlas of vulnerability and impact of climate change in the Gran Chaco Americano, Aug 2014. Development, Participation and Citizenship Institute, National University of Formosa, Cordillera Foundation, Norwegian Ministry of Foreign Affairs, Government of Spain, Gran Chaco American Center of Excellence, Asunción, Paraguay

Seo SN, Mendelsohn RO (2008) An analysis of crop choice: adapting to climate change in South American farms. Ecol Eco 67(1):109–116

Tubiello FN, Rosenzweig C (2008) Developing climate change impact metrics for agriculture. Int Assess 8(1)

UNESCO, PHI-LAC (2010) Atlas of arid zones of Latin America and the Caribbean

University of Cambridge (2014) Climate change: implications for agriculture. Climate everyone's business. Business for social responsibility (BSR), June 2014. European Climate Foundation (ECF), University of Cambridge's Judge Business J School (UCJBS), Institute for Sustainable Leadership (CISL)

Vega García H (2008) Climate variability and change, poverty and migratory challenges: the Central American case. In: III international colloquium on migration and development. International migrations: the challenges of exclusion and inequality for citizens in globalization. New analytical and conceptual perspectives, Heredia, Costa Rica, 4–6 Dec 2008

Vivero Pol JL (2013) Seasonal hunger in Guatemala: analysis of variables, monitoring methodology and public policy proposals to combat it. In: Final consulting report: the action plan against seasonal hunger within the framework of the zero hunger pact plan. Analysis of variables and intervention proposals. Support program for the national food and nutritional security policy of Guatemala and its strategic plan. EU, July 2013. Government of Guatemala, Catholic University of Louvain

World Bank (2014) World development indicators: electricity production from hydroelectric sources. Washington, DC

World Food Programme (2002) Standardized food and livelihood assessment in support of the Central American regional protracted relief and recovery operation. Regional Bureau for Latin America and the Caribbean

WRI (2016) As clouds head for the poles, time to prepare for food and water shocks. http://www.wri.org/blog/2016/07/clouds-head-poles-time-prepare-food-and-water-shocks?utm_campaign=wridigest&utm_source=wridigest-2016-07-26&utm_medium=email&utm_content=learnmore

Chapter 4
Energy and Climate Change: Challenges for Low-Carbon Development

Abstract This chapter describes the impact of the energy sector in climate change, both from the point of view of its importance in the international negotiation process to find a sustainable solution to the problem and its responsibility for the increase in atmospheric concentrations of greenhouse gases (GHGs) and the opportunities for adaptation and mitigation presented by the sector. On the other hand, some challenges that arise in the implementation of policies and measures to address the problem are analyzed.

Keywords Energy sector · Vulnerability · Adaptation · Mitigation · Policies and measures · Historical responsibility

---

Leónidas Osvaldo Girardin. Researcher at the National Council of Scientific and Technological Research of the Argentine Republic (CONICET). Researcher and former Director of the Environment and Development Program of the Bariloche Foundation (FB). Full Professor of the Economics and Environment Area of the Department of Applied Sciences and Technology (DCAyT) of the National University of Moreno (UNM).

Originally based on the recording of a presentation made by the author at the Buenos Aires headquarters of the Fundación Patagonia Tercer Milenio on August 9, 2012. Girardin (2012).

L. O. Girardin, *Socioeconomic and Geopolitical Aspects of Global Climate Change*, The Latin American Studies Book Series, https://doi.org/10.1007/978-3-031-53246-7_4

4.1 The Context of International Negotiation and Heterogeneity

The international negotiation on climate change is in a moment of "quagmire", as it does not show the results expected from it. Both at the beginning of the 1990s, when the United Nations Framework Convention on Climate Change (UNFCCC)[1] was signed, and at the end of the same decade, when the Kyoto Protocol (KP)[2] was subscribed. And also at the end of 2015, when the Paris Agreement (PA)[3] was endorsed, other expectations were raised. It has been known for some time that the limits on greenhouse gas (GHG) emissions set for the so-called "first commitment period" of the KP (2008–2012), which were judged at the time to be extremely moderate,[4] would not be sufficient to avoid harmful effects on the earth's climate.[5] However, it was later found that these modest commitments have also not been fully met by some of the main actors involved, mainly the USA.[6] This situation worsened after the Doha meeting (December 2012),[7] when some countries that had committed to quantified targets in the previous period[8] withdrew from the KP's "second commitment period" (2013–2020). Something similar is happening with the PA, in terms of the efforts that are still needed to ensure the targets that were proposed in the PA.[9] The agreements are very laborious to devise and, at the same time, insufficient to be reassured about the consequences that will probably have to be faced in the future.[10]

Thus, it is almost certain that the objectives of any of the three instruments signed in the framework of the negotiations will not be met in this way: neither those of the Convention, nor those of the Protocol, nor those of the Agreement. Thus, the need to deepen the commitments assumed arises, while there is a very strong inertia in the negotiation toward the lack of concrete actions. This situation is mainly explained by the interests on which the countries' positions are based and also by other structural

---

1 UNFCCC (1992). The United Nations Framework Convention on Climate Change was signed in June 1992 at the United Nations Conference on Environment and Development, held in Rio de Janeiro on that date, and entered into force on March 21, 1994.

2 UNFCCC (1997). The Kyoto Protocol was signed in December 1997, within the framework of the COP-3, and entered into force on February 16, 2005.

3 UNFCCC (2015). The Paris Agreement was initialed at COP-21 on December 12, 2015.

4 See, for example, Barros et al. (2005).

5 See, mainly, the documents related to the IPCC-Fifth Assessment Report 2013–2014 (5AR) y Fourth Assessment Report 2007 (4AR). IPCC (2007a, b, c, 2013, 2014a, b); www.ipcc.ch.

6 See http://unfccc.int/ghg_data/kp_data_unfccc/items/4357.php.

7 UNFCCC (2012). 18° Conference of the Parties to the United Nations Framework Convention on Climate Change (COP-18).

8 Japan, Russian Federation, Canada, New Zealand. See the Doha Amendment to the Kyoto Protocol. UNFCCC (2012).

9 Limit the global average temperature increase to below 2 °C from pre-industrial levels and commit the Parties to continue to work to make additional efforts to ensure that the temperature increase does not exceed 1.5 °C. UNFCCC (2015). See also UNEP (2017) and IPCC (2018).

10 IPCC (2018).

factors, which have to do with the specific characteristics of some of the main sectors responsible for greenhouse gas (GHG) emissions, which are the main human cause of interference with the climate. Among these sectors, the energy sector plays a predominant role.

In this context, in the so-called "Cancun Agreements"[11] of almost a decade ago, it is necessary to think about "low-carbon development", both in the paragraphs corresponding to the so-called "Shared Vision"[12] and in those corresponding to the mitigation measures.[13]

It is important to note that, regardless of the success or failure of measures taken with regard to the prevention of climate change and the possibility of adapting to its expected consequences, the results of international negotiations in this field will, sooner rather than later, have an impact on the modes of production and consumption patterns of the various societies in the different countries and, consequently, on the various economic sectors and activities with which they are related.[14] This influence occurs because, regardless of whether or not the expected impacts are verified, some measures are already being taken (mainly by those countries that have assumed some kind of quantitative commitment to limit and/or reduce their GHG emissions in the UNFCCC and the KP), and these measures will have a different impact depending on which economic sectors and/or activities they affect. Thus, strategies, policies, and measures aimed at achieving low-carbon development present a number of challenges in terms of assessing the real possibility of achieving any of these objectives in the short or medium term.[15]

At this point, it is essential to contextualize the problem of climate change. It is common to hear a supposedly ingenuous (or "naïve") version or account of problems related to the environment on a global scale (mainly those related to climate change), in the sense that these are problems that "affect all of humanity equally" or that they are issues related to things that "are going to happen to all of us" and, therefore, issues common to all the inhabitants of the planet. And, while this is partly true, there are certain considerations that need to be raised.[16]

---

[11] UNFCCC (2010). The Cancun Agreements were established within the framework of the 16th Conference of the Parties to the United Nations Framework Convention on Climate Change (COP-16), held in Cancun, Mexico between November 29 and December 10, 2010.

[12] Paragraph 6 of the Cancun Agreements: "(…) agrees that Parties should cooperate in achieving the peaking of global and national GHG emissions as soon as possible, recognizing that the time frame for achieving this will be longer in developing countries and being in mind that social and economic development and poverty eradication are the overriding priorities of developing countries and that a low-carbon development strategy is indispensable to sustainable development; (…)". See UNFCCC (2010).

[13] Paragraph 10 of the Cancun Agreements: "(…) Realizes that addressing climate change requires a paradigm shift towards building a low-carbon society that offers substantial opportunities and ensures continued high growth and sustainable development, based on innovative technologies and more sustainable production, consumption and lifestyles, while ensuring a just transition of the workforce that creates decent work and quality jobs (…)". See UNFCCC (2010).

[14] Girardin (2000, 2013).

[15] Girardin (2000, 2013).

[16] Girardin (2000, 2013).

Firstly, the issue of climate change is a global problem, because it is true that it will affect the planet as a whole. Moreover, it is perhaps (today) the largest global environmental problem, mainly since the late 1980s.[17] From that time onward, a series of empirical evidences began to take on greater relevance, showing that something was indeed happening to the climate, and that these transformations went beyond the natural changes that took place throughout the planet's history, and that, at the same time, they were strongly related to human activities.[18]

From a certain point of view it is undeniable that the problem is planetary, since all the inhabitants of the planet, in some way, are going to suffer in greater or lesser magnitude the consequences, and in that sense, we all have some degree of responsibility to take measures to prevent these changes and interferences, on the climate, that originate in human activities. However, it is also strictly true that these expected consequences are not going to impact all the inhabitants of the different regions of the world in the same way and that the responsibilities for having reached this situation are not the same for all the actors either.[19]

From this situation, the first concept that is fundamental to highlight, which crosses over this whole issue, is the question of heterogeneity. Heterogeneity is presented not only from the geographical point of view (related to which regions are expected to suffer most from these impacts), but also from the social point of view (which social groups are going to be most affected) and economic point of view (which activities are going to be affected); since the impacts expected are not necessarily equivalent in all cases (we are talking about regions, countries, economic activities, or social groups linked to all these elements), nor are the same effects expected everywhere. A concrete example of this heterogeneity in expected impacts is the case of Argentina itself.[20]

Argentina is a very large country that has a series of diverse climates in its territory and therefore different regional specializations in terms of economic activities and different social groups associated with them. This situation implies a high degree of heterogeneity, both in terms of the potential threats and risks in terms of the climate impacts expected in each region, as well as in terms of the potential impact on these economic activities. With regard to some of the effects observed in recent years, which can be attributed to climate change, in a country like Argentina, the issue of observed changes in rainfall can be mentioned as an example. On this point, two very different phenomena are evident, according to which region of the country is

[17] It is interesting to remember that, before the end of the *Cold War*, the main global environmental problem that was expected and the main fear was that a possible war between the USA and the former Soviet Union would bring about a "nuclear winter". These theories were based on studies from the early 1980s by Crutzen and Birks (1982), Turco et al. (1983), and Aleksandrov and Stenchikov (1983), which discussed the effects on the earth's climate of smoke and particles from a possible nuclear war between the two great superpowers of those times.

[18] On early awareness of the heterogeneity of the geographical distribution of impacts, see IPCC (1990). Regarding the heterogeneity of "responsibility", reference is made mainly to the concept of "common but differentiated responsibilities" referred to in Article 3 of the UNFCCC (1992).

[19] Girardin (2000, 2013).

[20] Girardin (2000, 2013, 2017b) and ECLAC (2014).

taken. On the one hand, there is an increase in rainfall in the Pampean and Chaco plains that has been occurring in the last 70 or 80 years and, on the other hand, an exacerbation of situations of water stress and a certain deficit of rainfall in the area of Cuyo and Comahue. In other words, within the same country there are two totally different phenomena, which means that geographically the impact of climate change will be very heterogeneous.[21]

From the social point of view, the impacts will also be very heterogeneous. And in this aspect the concept of vulnerability appears in all its importance (which is also fundamental in the analysis) since it crosses all the elements related to this issue. Vulnerability is defined in the sense of how different social groups and different ecosystems are prepared to face those changes that go beyond the average. Those changes, due to their magnitude or frequency, do not represent what was statistically expected to happen in comparison with what happened in previous years.[22]

The concept of vulnerability has two fundamental components: on the one hand, the magnitude of the impact and, on the other hand, the capacity of that community or ecosystem to react to that impact. The vulnerability of an impoverished population with poor access to health or minimal infrastructure will not be the same as that of another social group with better conditions for coping with climate change. As can be seen, there will also be varying degrees of heterogeneity between different sectors in terms of their characteristics and response capacity, which will make them more or less vulnerable to the various impacts that may be expected.[23]

In the case of the degree of responsibility of the different countries and regions if the current situation has been reached, the heterogeneity is also very strong and plays a fundamental role in the discussion of this problem. However, it is not always clearly highlighted. Although, as far as the subject of climate change is concerned, reference is generally made mainly to issues related to GHG emissions; in reality this is a clearly cumulative process, in which what matters is not only current GHG emissions, but mainly the atmospheric concentrations of these gases. These atmospheric concentrations depend fundamentally on two things: past emissions and the degree of permanence in the atmosphere of each of these gases (which is not necessarily the same for all of them).[24]

In this regard, it is important to emphasize that carbon dioxide (CO_2) is the main greenhouse gas for whose emissions human activities have a major responsibility. At the same time, the main source of emission of this gas is the massive burning of fossil fuels. Bearing in mind these considerations and the fact that the time carbon dioxide

---

[21] See the results of both the Second and Third National Communication of the Argentine Government to the United Nations Framework Convention on Climate Change (2CN and 3CN) and their complementary documents. Government of the Argentine Republic (2007, 2015), available at www.ambiente.gob.ar y en www.unfccc.int. See also ECLAC (2014) and Girardin (2017b).

[22] See the reports produced by the IPCC Work Group 2 (WGII). IPCC (2007b, 2014a). Available at www.ipcc.ch.

[23] Girardin (2000, 2013).

[24] See, for example, IPCC (1994, 1995, 2007a, 2013). Available at www.ipcc.ch.

remains in the atmosphere is very long (it can remain in the atmosphere for around 50–200 years),[25] it is easier to understand why there is a large consensus among experts, supported by empirical evidence, that puts the Industrial Revolution as the starting point of the phenomenon of climate change originating in human activities.[26] With the Industrial Revolution, the human species begins to massively burn fossil fuels. At this point, the responsibility for having reached the current situation of saturation in the atmospheric capacity to absorb these gases without interfering with the balances that historically existed naturally in the earth's climate is clearly different among those countries that first agreed massively to the burning of fossil fuels: first wood, then coal, and finally oil, its derivatives, and natural gas. In this group of countries that acceded before others to these styles of production, we find the USA, the countries of Western Europe, Eastern Europe, Japan, and other developed countries of the Asia–Pacific (Australia, New Zealand). Their responsibility for having reached the current situation is clearly greater than that of the countries that later acceded to the process of the Industrial Revolution and the growth model based on energy available in large quantities and at low prices.[27]

Why is the issue of responsibility important in international negotiations? Because the burning of fossil fuels is a fundamental factor (although not the only one) in the emission of GHGs, this situation must be analyzed taking into account that the opportunities to meet development needs in the so-called "emerging" countries are increasing. That is, countries with large portions of their population with quality of life standards are still very low and that, surely in the future, will incorporate higher standards of comfort and quality of life. This process will necessarily result in these countries emitting much more than they do now, even though they have more efficient consumption and production patterns than those used for their development by the countries that are now developed.[28] And that is important in the international negotiation on climate change, because the final result is not the same when you talk in relative terms as when you talk in absolute terms: when you talk about total tons emitted or when you talk about tons emitted per capita. It cannot be ignored that countries like China and India have a fundamental weight in absolute terms, because they represent more than a third of the world's population. But, on the other hand, it is also legitimate to ask why an inhabitant of China or India does not have the same right to use that portion of the atmosphere's GHG processing capacity that the inhabitants of more developed countries make and have made use of.[29]

---

[25] See, for example, IPCC (1994), Table 3. www.ipcc.ch.

[26] IPCC (1990, 1992). www.unfccc.int.

[27] Girardin (2000, 2013), GEA (2012), and Baumert et al. (2005).

[28] See, for example WBCSD (2010).

[29] See Lipietz (1995) and Girardin (2000, 2013).

4.2 Equity and Efficiency. Future and Current Generations

The other important issue to address, in the analysis of both socioeconomic and political aspects related to climate change, is that of decision-making despite the prevailing uncertainty. Why? Because the ultimate consequence of this phenomenon can have such an impact (some scenarios pose catastrophic situations)[30] that it is very difficult to measure in economic terms. Therefore, the most rational thing to do would be to act on the causes known today, trying to avoid certain effects in the future. This is recognized in the UNFCCC itself, which sets out the "precautionary principle".[31] But it turns out that this logic is absolutely contrary to economic logic. Economic logic says that human beings have a preference to consume today, beyond the fact that something may actually happen in the future. And it translates that "time preference for today's consumption" into the use of a "discount rate". In other words, everything that one consumes (or suffers) in the future is applied a discount rate that ultimately makes everything that happens in 30 or 40 years' worth much less than what happens today.[32]

Obviously, if decisions are made on the basis of applying these economic criteria, the use of a discount rate (e.g., 10% or 12%) may lead to the consideration that any potential disaster that may occur in 100 years' time is not a problem worth solving through an investment that means devoting a large amount of financial resources today. Thus, what the application of this traditional cost–benefit analysis raises, from the economic point of view, is that it will never seem a good time to apply preventive actions and, what is worse, is that this behavior does not generate any theoretical problem for decision-making within that approach.[33]

So, this is one of the reasons why decision-makers have problems when they talk about very long-term issues. Why? Because there are two typical mistakes that can be made in the decision-making process: (a) you do not make investments today for something that could be a catastrophe in the future; or (b) you make a decision to make an economic effort today for something that may not happen tomorrow. Of these two mistakes, policymakers are under strong pressure to make the first one. Thus, except in exceptional cases, it is a matter of minimizing spending today, not making sacrifices and not passing on the cost of prevention to those who vote for them today. Tomorrow, it will be another day, and if the situation explodes, another one will take over. To avoid this type of behavior and to tend to proactive action in favor of prevention, a major effort must be made to raise awareness, not only among decision-makers but fundamentally in society, so that the latter will be the one to put pressure on decision-makers and policymakers. Policymakers are forced to make these decisions in a context of strong uncertainty, so that they often do not

---

[30] IPCC (2013, 2014a).

[31] CMNUCC (1992).

[32] This was discussed in numerous papers devoted to the relationship between economics and climate change. In the case of Argentina, see, for example, ECLAC (2014) and Girardin (2017a). In the case extended to the global level, see Stern (2006).

[33] Nordhaus (2007), Stern (2006), ECLAC (2014), and Girardin (2017a).

depend exclusively on scientific knowledge but, fundamentally, on the political will of decision-makers and their perception of society's demands.[34]

Another issue that is also important to contextualize is that mitigation measures[35] could mean a very costly sacrifice in terms of resources at the time of decision-making, especially in countries where economic and/or financial means are not sufficient. And while, in terms of commitments made and responsibility, the country is obliged to carry out mitigation measures, it is also true that the time scale on which certain measures must be taken is very different. Why? Because with regard to adaptation, it can be seen that in many cases the impacts attributable to climate change and/or climate variability are already occurring and are observable (as in the case of the El Niño and La Niña phenomena, for example, where these effects have been very significant).[36] However, measures taken to reduce and/or limit GHG emissions may have observable, measurable, and verifiable effects only over a relatively long period of time.[37]

Measures to adapt to climate change or climate variability, in general, have to be applied immediately (often in response to the emergency), so that their relevant periods are much shorter than the periods in which mitigation policies begin to be effective (which generally depend on the degree of permanence in the atmosphere and the impact on climate change of the GHG on which they are acted upon). This means that, in the latter case, perhaps a mitigation measure taken today will have an impact several decades from now. However, it is also important to stress that such a measure must be taken at some point, to avoid the accumulation of effects.[38]

Thus: sometimes decisions can result in synergistic actions between adaptation and mitigation; and sometimes conflicting. Faced with this dilemma, another instance will be opened in which methods and objectives will have to be prioritized.

Returning to the issue of responsibility for having reached the current situation, it is very common to read or hear that the best thing to do is to reduce a ton of CO_2 where it is cheaper to reduce it.[39] But this criterion of "global efficiency", applied worldwide, is not necessarily true since income distribution generally has some influence on the value of goods. And in terms of environmental goods, the same is true: land in Africa or South America does not have to be of lower quality or less suitable for agricultural use than in the USA, but it is certainly worth less.[40] That is why when we talk about "global efficiency" perhaps what we are really talking about is that the goods in the places where the poor live are worth less. So it is cheaper to reduce an impact on the poor than on the rich. This point is no less important, since the search for so-called "global efficiency" is an argument generally made

34 For issues related to decision-making in contexts of uncertainty, see Funtowicz and Ravetz (1993).

35 In terms of the "jargon" of climate change, mitigation is carrying out activities that reduce the emission of gases that produce the greenhouse effect.

36 See ECLAC (2014) and Gobierno de la República Argentina (2007, 2015).

37 Girardin (2000, 2013).

38 On the importance of taking mitigation measures as soon as possible see UNEP (2017).

39 See Girardin (2000, 2013).

40 See Girardin (2000, 2013).

by representatives of the most developed countries at international climate change negotiation meetings.[41]

The question of equity is aimed at the negotiation that the less developed countries and fundamentally the more populated ones, such as China and India, pose. If every Chinese had the level of energy consumption that every American has, it would undoubtedly be a very great pressure for the whole system, but why cannot they aspire to have it? In any case, it is necessary to arrive at a certain rule of "contraction and convergence" that implies that those who are squandering resources consume less and those who have not yet reached standards of living appropriate to the conditions of justice and dignity of each era can increase their consumption.[42]

Argentina emits about 0.7% of total world emissions.[43] In other words, although Argentina is among the top 30 emitters worldwide, this does not make it a decisive player in the process of preventing the harmful effects of climate change, even if it strictly complies with the commitments it takes on. However, this does not mean that Argentina's economic activities and ecosystems do not suffer the consequences of climate variability and what so far are evident as permanent changes in the climate. It will then be necessary to determine the priorities in terms of territory involved, the systems (natural or social) impacted, or people potentially affected. It would not be surprising if the analysis of all these variables resulted in the fact that the higher cost of not acting for a country like Argentina is more linked to the lack of action in adaptation policies than to the same behavior in mitigation policies.[44]

A crucially important issue is related to the economic valuation of potential impacts and measures. Generally speaking, the first question decision-maker asks is how much he or she will have to invest. Then, to make decisions often, previously, it is necessary to make economic assessments. There are methods that are based on the concept of "willingness to pay" (how much people are willing to pay to have this or that done or not done). And this concept is very much influenced by income level, since the expectation is that no one will be able to show a willingness to pay that goes beyond their income level.[45] If this criterion is applied to decide whether or not a decision will be made, it will influence much more the decisions of those who access higher income levels because the latter are willing to pay much more than those who can only reach lower income levels. So we have to be careful because sometimes trying to put a value on the economic impact on the environment does not necessarily mean that we are doing things right. But, in fact, there must be some indicator that influences the decision-making about whether something should be done or not.[46]

From the communication point of view, there is another issue that has to do with a certain "naturalization" of climate change, in the sense that the impacts of climate

[41] See Argawal and Narain (1991), Lipietz (1995), Bhaskar (1995), and Girardin (2000, 2013).

[42] Meyer (2000) and Girardin (2000, 2013).

[43] Gobierno de la República Argentina (2016).

[44] ECLAC/CEPAL (2014) and Girardin (2013, 2017a).

[45] Azqueta Oyarzún (1994) and Martízez Alier (1995).

[46] Azqueta Oyarzún (1994) and Martínez Alier (1995).

change appear as something natural; a situation in which nothing can be done. While these extreme weather events cannot be prevented from occurring, what can be done is to minimize the loss of life and resources by applying prevention strategies.[47] Generally, issues related to climate change appear in the media when there is a catastrophe with numerous victims and economic losses, or when a meeting is approaching that is expected to have some relevance, to disappear from the informative agenda the rest of the time. This combination of "informational sensationalism" and "naturalization of consequences", which starts from the premise that climate change is a natural and inevitable phenomenon, leads to resignation about how to face it and ends up in a discourse that leads to inaction.[48]

In this sense, there is enough to say in favor of having a more "sociological" vision of these phenomena taken, generally as "natural". Often, the effects of extreme events caused by climate variability or change are mounted on a system of inequalities, inequities, or problems that are pre-existing. At other times, human actions exacerbate problems that later become unmanageable in the face of extreme climate events.[49] One example is the flooding of Santa Fe in 2003.[50] Is that the fault of the weather? The answer, clearly, is: no. The 2003 floods in Santa Fe have both natural and man-made causes. The origin was intense rainfall, over a saturated basin (among other things, due to inadequate land use), the absence of prevention measures, and the lack of completion of the necessary infrastructure works.[51] But many times the combination of inadequate information "catastrophism" and "naturalization" of the problems leads to inaction and to think that nothing can be done.

It is generally said that those most vulnerable to climate change are the poorest. This is true, but because the poorest suffer from what is often called "multiple exposure": not only are they the most vulnerable to extreme events of climate variability and change, but also to the consequences of global economic processes, to lack of access to energy, food and sanitation, to diseases and epidemics; and, in general, to almost all sudden or unexpected changes they have to face.[52] Climate change is likely to expand the asymmetries between those who are better prepared to cope, in some way, with the expected impacts, and those who cannot. So, will not an integrated development policy be a measure of adaptation to climate change? It certainly is. A more informed society with better health conditions, better nutrition, better access to infrastructure, and various containment systems will surely be in a better position to deal with any phenomenon that exceeds the average.[53]

[47] Girardin (2000, 2013) and Herzer (1990).

[48] Girardin (2000, 2013) and Herzer (1990).

[49] Herzer (1990) and Girardin (2013).

[50] UNL (2003).

[51] UNL (2003).

[52] Postigo et al. (2013).

[53] Girardin (2000, 2013) and Postigo et al. (2013).

In relation to this issue, it is often argued that migration (internal and external) is linked to climate change.[54] However, in most cases, the main cause of such migrations is not so much the effect of climate change or variability, but rather the conditions of poverty. Yes, it may be that climate change or variability generates the migration of some specific communities that are associated with certain cultural practices. But in general, if it is necessary to migrate because the climate has changed, the conditions of poverty must be analyzed in such a way that it is evident that there is no alternative activity that allows survival. This is another case among which the naturalization of climate issues hides other issues related to the social, economic, technological, and political,[55] where the issue of climate change in migration is key is in the case of the potential rise in mean sea level.[56]

In the case of a food producing and exporting country like Argentina the issue of food security put at risk by climate change or variability would have no reason to become a critical issue,[57] but in other regions (the Middle East, Central America, and some other Latin American countries) it is an issue that may become more and more important.[58]

4.3 Energy and Climate Change

Speaking of the relationship between the energy sector and climate change and based on hard energy data, commercial energy consumption by region is still higher in developed countries than in developing countries, beyond the trends expected in the future.[59] What is shown in Graph 4.1 is actually the correlation that the situation described above has on GHG emissions, particularly in this case on CO_2 ones.

This graph shows that while developing countries have increased their share of total global CO_2 emissions (in this case in the energy and Industrial Process sectors), they have not necessarily done so in the same proportion as China, India, and some Southeast Asian countries. It is clear that Latin America, for example, has maintained practically the same proportion on global emissions. In the case of Africa, the

[54] The International Organization for Migration (IOM) is carrying out the MECLEP project (Migration; Environment and Climate Change: Empirical Data for Policy Making), which has a portal that presents detailed information on this topic. See https://www.iom.int/es/migracion-y-cambio-climatico and http://www.environmentalmigration.iom.int/es.

[55] Girardin (2000, 2013).

[56] IPCC (2014b).

[57] In any case, it is more an issue that should be located as a distributive failure, rather than a productive one, as there would be enough resources to feed its own population several times over but this does not guarantee that all those who need food have access to it.

[58] For example, the areas of the fourth, fifth, and sixth regions of Chile, that is to say, the wine and agricultural areas par excellence of that country, are going to be subject to water stress. So they will have to expand the border to the south or to the north taking into account, among other things, the suitability of the soil. CEPAL (2012).

[59] GEA (2012), IEA (2007), Olivier et al. (2016), FAO (2008), and USEIA (2007).

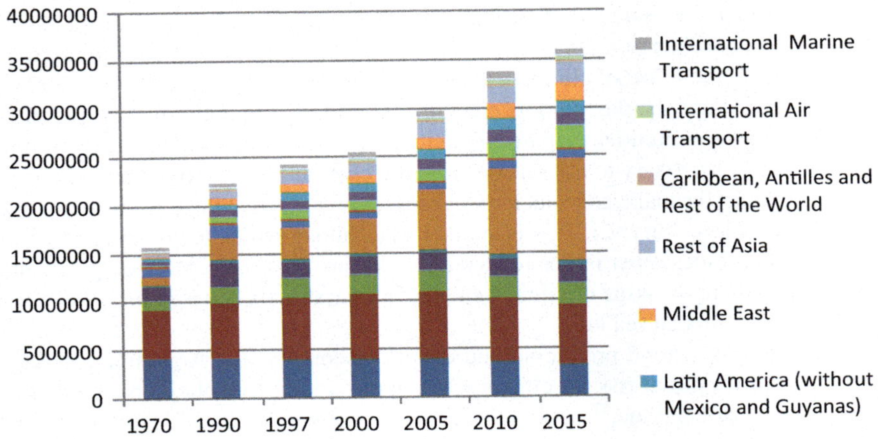

Graph 4.1 Participation of CO_2 emissions corresponding to the burning of fossil fuels and industrial processes. Gg of CO_2/year. *Source* Own elaboration based on Olivier et al. (2016)

behavior shown in the graph represents mostly the oil countries of the north of the continent and South Africa, which is the most developed country of all of them as well as a large producer of coal.[60]

In terms of CO_2 emissions, developed countries that are part of the Organization for Economic Cooperation and Development (OECD)[61] which brings together most of the countries of Eastern and Western Europe,[62] developed countries of Asia–Pacific Region (Japan, Australia, New Zealand), USA, Canada, and the new members of the entity (Mexico, Chile, and South Korea), together with the Russian Federation, Ukraine, and other nations of the former USSR and Eastern Europe (except those now included within the OECD), accounted for more than half of the CO_2 emissions from the energy sector and Industrial Processes until 2005. This participation has been reduced over time from 81% of the world total in 1970; to 71.3% in 1990; 62% in 2000; 55% in 2005; 47% in 2010; and 42% in 2015. At the time of the signing

[60] GEA (2012), IEA (2007), Olivier et al. (2016), FAO (2008), and USEIA (2007).

[61] The OECD currently includes the following countries: Australia, Austria, Belgium, Canada, Chile, Czech Republic, Denmark, Estonia, Finland, France, Germany, Greece, Hungary, Iceland, Ireland, Israel, Italy, Japan, South Korea, Latvia, Lithuania, Luxembourg, Mexico, Netherlands, New Zealand, Norway, Poland, Portugal, Slovak Republic, Slovenia, Spain, Sweden, Switzerland, Turkey, the UK, and the USA.

[62] The European Union (EU27) consists of the following countries: Austria, Belgium, Bulgaria, Cyprus, Czech Republic, Denmark, Estonia, Finland, France, Germany, Greece, Hungary, Ireland, Italy, Latvia, Lithuania, Luxembourg, Malta, Netherlands, Poland, Portugal, Romania, Slovak Republic, Slovenia, Spain, Sweden, and the UK. The EU28 adds Croatia. Although almost all the members of the EU28 are also members of the OECD, Bulgaria, Croatia, Romania, Malta, and Cyprus are not members of the latter. Likewise, Norway, Turkey, Iceland, and Switzerland, although located in European territory, are not part of the EU28.

of the KP, the participation of this group of nations over the world total of CO_2 emissions in these sectors mentioned amounted to 63%.[63]

The fall in the share of "developed world" emissions in the global total was offset by a notable increase in China's share of that total, which went from just under 6% of the total in 1970 to just over 10% in 1990; 14% in 2000; 20.7% in 2005; 26.7% in 2010; and almost 30% of the world total in 2015. According to this source, the joint participation of China plus India and the rest of Southeast Asia in 2015 (around 42% of the total) was equal to that of all the "developed" nations.

However, for the rest of the regions, the evolution of their emissions was much less "explosive". Thus, for the rest of Europe that is not part of the OECD (mainly some countries of the former Yugoslavia, Moldova, Albania), the participation in the total world emissions, between 1970 and 2015, oscillated around 0.3% worldwide, with a peak of 0.5% in 1990.

In the case of the Middle East, there has been a significant growth in its share of total emissions (mainly explained by the increase in oil production), rising from 1.45% in 1970 to over 5% of the total in 2015.

Africa shows a fairly stable behavior in the share of its emissions throughout the series, going from representing 2.27% of the world total in 1970 to stabilizing around 3.45% in the last 15 years.

The share of Latin American emissions (excluding Mexico, the Guyanas, and some smaller Caribbean islands) in the world total also shows a stable trend. In 1970 they represented 2.4% of the world total (slightly more than 3% if Mexico is added), and over the last 20 years, this proportion has stabilized at around 3.5% worldwide (slightly less than 5% of the total if Mexico is included).

The rest of the regions not included in this classification (*The Guyanas*, other small islands in the Caribbean, Polynesia, Micronesia, and other smaller islands in the Atlantic, Pacific, and Indian Oceans) have a share that, over the entire period analyzed, barely exceeds 0.2% of the world total; much less than the combined share of emissions from International Air Transport and International Maritime Transport, which over the same period represents about 3.2–3.3% worldwide, except in 1990, when their share was only 2.8%.

Graph 4.2 represents a flowchart of global GHG emissions due to anthropogenic causes.

It arises from the flowchart that the main GHG is carbon dioxide, CO_2 (77% of total emissions) whose largest source of emissions is the energy sector, mainly the burning of fossil fuels (57.5% of the global total), in the so-called energy industries, mainly electricity generation and refineries (24.6% of the total), in transport (13.5%), in energy uses in the industrial and construction sector (10.4%) and energy consumption in commercial, public, and residential buildings, and other sectors such as fishing and agriculture (9.0%). Another important source of CO_2 emissions is the land use, land use change, and forestry (LULUCF) sector, which represents 18.2% of total global GHG emissions.

[63] Olivier et al. (2016).

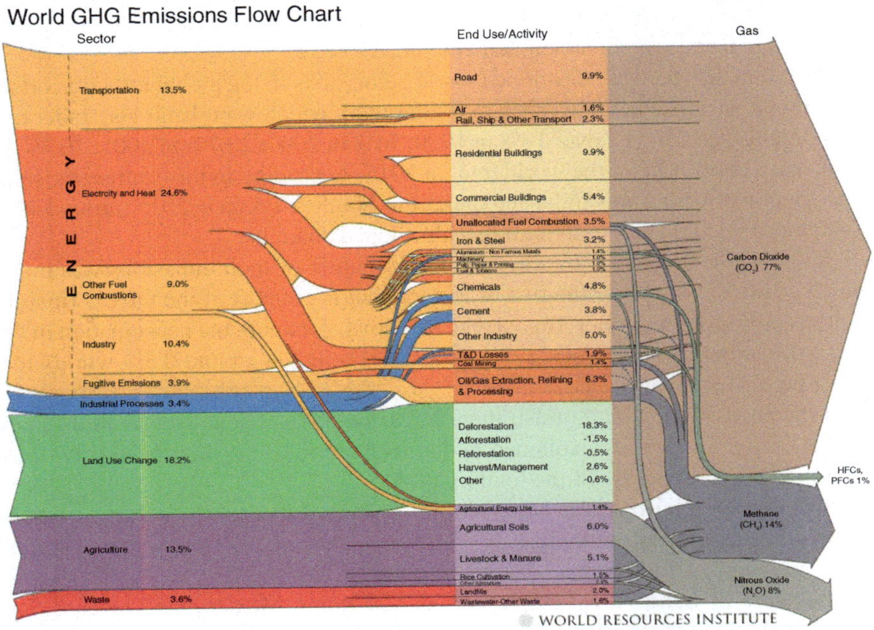

Graph 4.2 GHG emissions by sector and gas. *Source* WRI-CAIT (http://cait.wri.org/)

The second GHG in order of importance is methane, CH_4, which represents 14% and is mainly originated in the agricultural activity (Enteric Fermentation of Cattle, Handling of Manure, Rice fields), in the waste management sector (both in Urban Solid Waste Management, mainly in Landfills, as in Residential and Industrial Wastewater) and to a lesser extent, both in the form of Fugitive Emissions in the Production of Fossil Fuels (oil, natural gas, coal), as well as in emissions linked to Industrial Processes (mainly in the Petrochemical Sector).

In third place of importance (8% of the global total) are the emissions of nitrous oxide, N_2O, which is linked to agricultural soil use, fertilizer application, and manure management and to a lesser extent in the energy, waste management, and Industrial Processes sectors.

Graph 4.2 focuses on GHG emissions based on the activity that generates them, but without distinguishing by country and/or region. But when the analysis focuses on countries' GHG emissions, we have that about 20% of the world's nations emit about 80% of the GHGs originating from human activities. In terms of identifying the actors, Graph 4.3 clearly identifies which countries are responsible for that 80% of emissions, at present. These countries have an important specific weight, and it would surely be better to start the discussion of an effective and sustainable climate agreement with them.

However, the responsibility for current emissions does not necessarily coincide with the "historical responsibility" for past emissions and, consequently, for

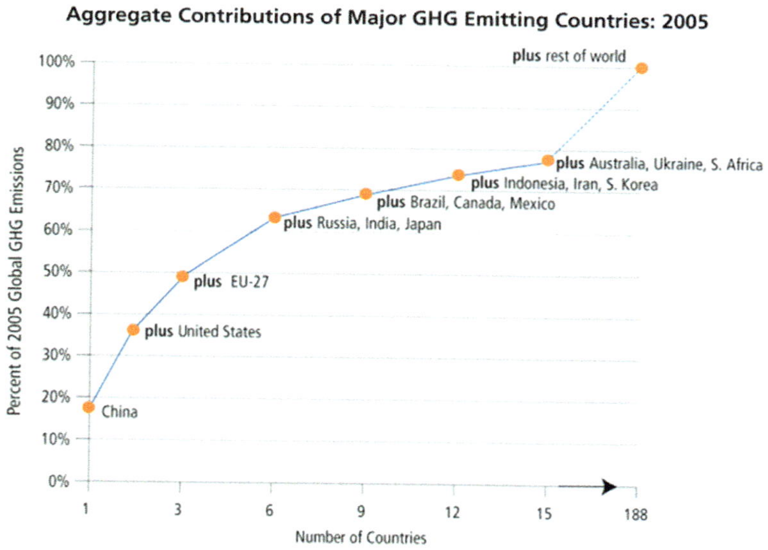

Graph 4.3 GHG emissions by country/region accumulated from highest to lowest (2005). *Source* Baumert et al. (2005)

current atmospheric concentrations of GHGs, which have been accumulating since the "Industrial Revolution". Graph 4.4 illustrates this situation a little better than the one described above.

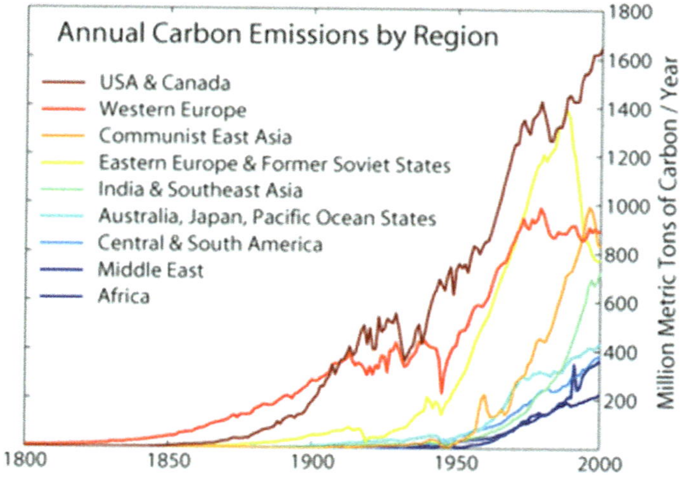

Graph 4.4 Historical accumulated emissions (1800–2000). *Source* GEA (2012)

As it can be seen, at the beginning of the series, the main emissions were from Western Europe (until the beginning of the twentieth century, with a very important participation of the UK), then there was a strong decrease during the Second World War and in the immediate aftermath (mainly due to the collapse of the industrial sector in the main countries of the region), a rebound in the post-war period and an uninterrupted growth in emissions until the 1970s. From that moment on, emissions stabilized, largely due to demographic stabilization, but in the initial period of that decade, also due to the 1973–1974 oil crisis.[64] This stabilization is also noticeable in the developed countries of Oceania and Asia–Pacific (predominantly Australia and Japan), although at much lower levels than in Europe.

In the case of emissions for the USA and Canada, the trajectory of these shows a constant increase with the only exceptions of the economic crisis of the years 1929–1930 and the oil crisis of 1973–1974. Unlike what happens with the group of countries analyzed in the previous paragraph, in this case there is no stabilization of emissions.

The nations of Eastern Europe and the former Soviet Union had a growing and very important participation, mainly from the middle of the last century until the years 1989–1990, when structural changes took place, both political and economic, which resulted in a significant fall in their levels of GHG emissions.

These three sets of countries have the highest cumulative historical values in the series and, consequently, the highest degree of responsibility for atmospheric GHG concentrations in the period analyzed.

From the last third of the previous century, we can see how the emissions of China, India, and Southeast Asia have grown significantly, making them the main players in terms of their GHG emission levels, but with a much smaller historical responsibility than the first three groups of countries mentioned. Likewise, the rest of the regions (South and Central America, the Middle East, and Africa) have a much lower historical responsibility than all the groups of countries mentioned above.

Graph 4.5 shows the historical evolution of world primary energy consumption by type of energy used over the period 1850–2000.

It can be observed that at the beginning, energy consumption was based on biomass (mainly wood) and how as time goes by "layers" of new types of energy are incorporated, starting with coal (which is still the fuel used in greater proportion until well into the twentieth century), followed by oil and its derivatives (which today is the most used), with a very important use of natural gas (mainly in recent years) and with a growing participation of nuclear energy and renewable energies from the last third of last century. What can also be clearly seen is that energy consumption at the beginning of the new millennium is five times higher than it was at the beginning of 1960.

Beyond what can be analyzed at a global level, there is a very large heterogeneity in the emission structures of the various countries, depending mainly on their productive structure and their energy matrix.

[64] GEA (2012).

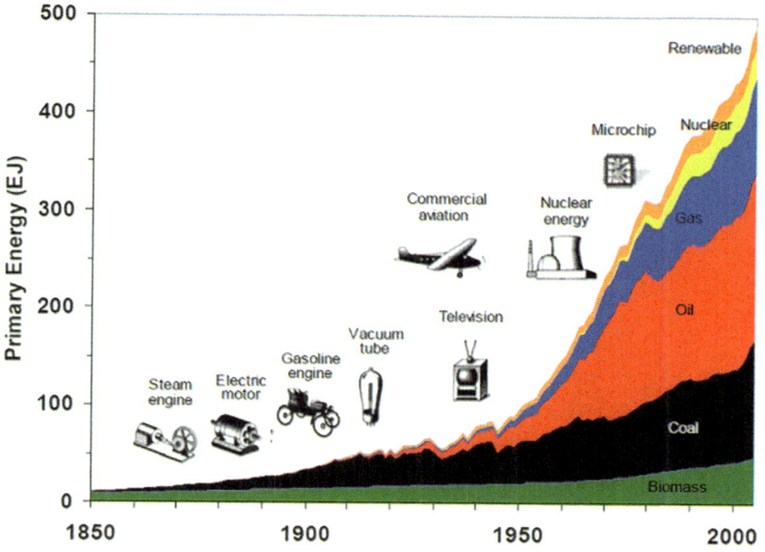

Graph 4.5 Global consumption of primary energy by type of energy and prevailing technologies at each moment in time. 1850–2000. *Source* GEA (2012)

As it can be seen, the GHG emission structures of the countries included in Graph 4.6 are very diverse, depending on their "national circumstances". This situation means that each nation will be able to identify different mitigation options in different sectors, and at the same time, it conditions in some way their positions in international negotiations on the issue in question. Although the vast majority of these countries are oil producers, and the main source of their GHG emissions (and therefore their mitigation opportunities) may be in the energy sector, this is not the case for all of them. While the energy component in China's GHG emissions is very important, in the case of Brazil a very significant proportion of emissions are related to the agricultural sector and, fundamentally, to land use change (deforestation). In the case of Argentina (at least at the beginning of this century), along with countries such as New Zealand and Uruguay, it shared the characteristic of having a very high component of GHG emissions from the agricultural sector that practically equaled those from the energy sector.[65] This structure of emissions in Argentina changes, as time goes by, because the cultivable area and the potential growth of heads of cattle, has a limit that is reached long before the limit of the increase of the energetic demand (mainly due to the increase of the population or the access to other energetic consumptions of this one). Thus, the percentage of emissions corresponding to the energy sector is growing over time as a proportion of Argentina's total emissions.[66]

[65] While in Argentina, emissions from the agricultural sector at the beginning of the twenty-first century were similar to those from the energy sector, in both Uruguay and New Zealand, emissions from agricultural activities were higher than those from energy activities.

[66] CEPAL (2014) and Girardin (2017c).

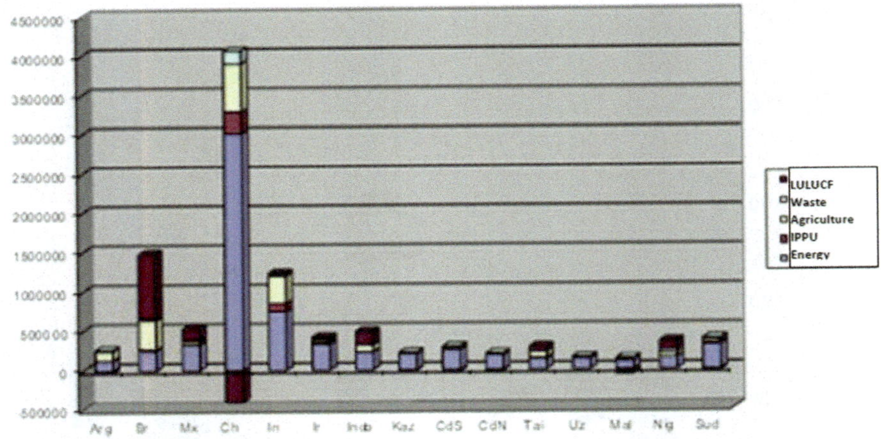

Graph 4.6 Structure of GHG emissions by sector of the 15 main non-Annex I emitting countries. Year 1994 in Gg of CO_2e. *Source* Girardin (2000) based on National Communications to UNFCCC of diverse countries: Argentina, Brazil, Mexico, China, India, Iran, Indonesia, Kazakhstan, South Korea, North Korea, Thailand, Uzbekistan, Malaysia, Nigeria, and South Africa

In this sense, below are some characteristics of the evolution of GHG emissions in Argentina. The latest National Inventory of Greenhouse Gas Emissions (INVGEI) available in Argentina corresponds to the year 2014 included in its Second Biennial Update Report (BUR2) to the UNFCCC.[67]

It arises as a result of the same that the total of GHG emissions recorded in the Argentine territory, in the year 2014, was 368,295 Gg of CO_2e. Of this total, the energy sector corresponds to 193,477 Gg of CO_2e, 52.5% of these emissions, making it the main emitting sector. If the shares of each sector are taken by type of gas, energy is responsible for 70% of total CO_2 emissions; 9.5% of the same in the case of CH_4; and 4% of N_2O emissions.

Between 1990 and 2012 the sector's emissions almost doubled from 103,464 to 193,477 Gg of CO_2e (an increase of 87%), from 35.23% of total emissions to 52.53%, between the beginning and the end of the series. This increase is higher than the average of all sectors, while the total GHG emissions of all sectors combined increased by 25.41% in the period analyzed.

Energy sector emitted 193,477 Gg of CO_2e emitted in 2014. About 94% of those emissions (182,299 Gg of CO_2e) correspond to fuel burning activities and the remaining 6% (11,178 Gg of CO_2e) to fugitive emissions. Of the emissions from fuel burning activities, 31.23% are explained by the burning of fuels in transport (56,929 Gg CO_2), mainly road transport (51,503 Gg CO_2e, more than 90% of the same); another 32% by the emissions from the burning of fuels in the energy industries (58,340 Gg CO_2), mainly in the production of electricity and heat (42,862 Gg CO_2e,

[67] Gobierno de la República Argentina (2017b). At the time of the publication of this article, it was also available the BUR3 that include the National GHG Inventory corresponding to 2016.

73.5% of that subtotal); and 25.3% to other sectors (46,119 Gg CO_2), among which the component of residential emissions is particularly relevant (28,415 Gg CO_2e, 62% of such emissions) and 11.47% corresponds to emissions from the burning of fuels in the manufacturing and construction industry (20,911 Gg CO_2), whose main component is the burning of fuels in the iron and steel industry (8538 Gg CO_2e, 41% of such emissions).

Graphs 4.7 and 4.8 illustrate what is commented in the preceding paragraphs.

The structure of Argentinean emissions shows a growing participation of the energy sector as a proportion of the total emissions and, at the same time, determines that the main mitigation potential is also related to strategies, policies, actions, and measures to be applied in that sector. An estimate of this potential in the long term is presented in Graph 4.9. While it can be seen that there is a potential to reduce and/or avoid GHG emissions in other sectors (waste management, agriculture, LULUCF, Industrial Processes), the main potential lies with the energy sector, comprising 85%

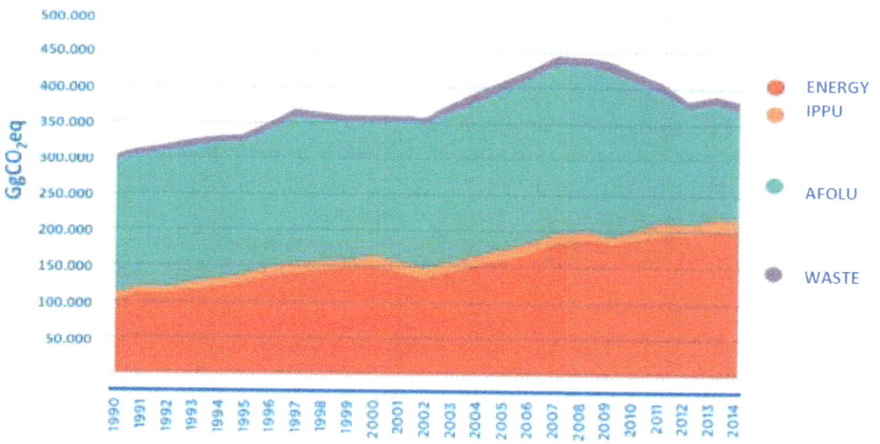

Graph 4.7 Argentina. Historical series of GHG emissions by sector (1990–2014) in Gg of CO_2e. *Source* Adapted from Gobierno de la República Argentina (2017b)

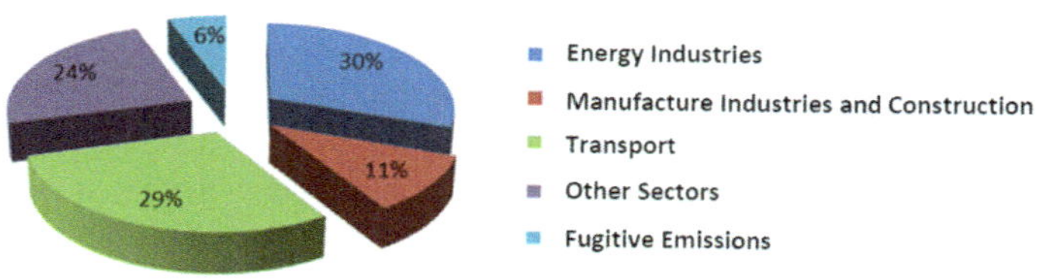

Graph 4.8 Argentina. Percentage composition of energy sector GHG emissions. *Source* Own elaboration based on BUR2 data

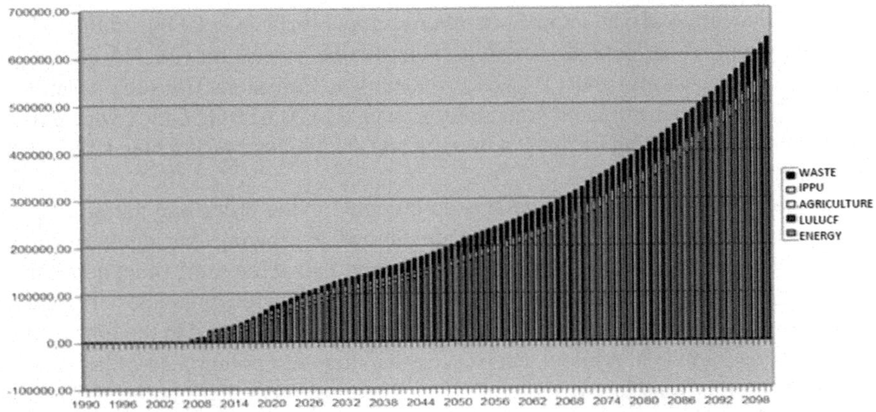

Graph 4.9 Net avoided emissions by sector for the period 1990–2100 in Gg of CO_2e. *Source* Adapted from CEPAL (2014) and Girardin (2017c)

of the potential reductions and/or limitations of emissions expected toward the end of the period analyzed.[68]

Although per capita emissions in Argentina are growing throughout the period covered by the studies,[69] there are options that allow us to think about limiting and/or reducing GHG emissions in the short and medium term, in the energy sector itself, tending to an improvement in the intensity of emissions as a result of improved energy intensity. Some of these measures identified are: changing the energy matrix in favor of less emitting alternatives; improving energy efficiency in transport; improving processes in industry; developing second generation biofuels; improving the efficiency of equipment used in homes; and improving passive efficiency in housing and public buildings.[70]

4.4 The "Inertia in Energy Consumption" and Future Perspectives According to the "Paris Agreement" and the Fulfillment of Its Goals

Graph 4.10 shows the evolution of world consumption of primary energy between 1973 and 2008.

It is interesting to note that energy consumption doubled (from just over 6 million TOE to over 12 million) and that beyond the changes brought about by the so-called

[68] CEPAL (2014) and Girardin (2017c).

[69] CEPAL (2014), Girardin (2017c), and Gobierno de la República Argentina (2007, 2016, 2017a, b).

[70] CEPAL (2014), Girardin (2017c), and Gobierno de la República Argentina (2007, 2016, 2017a, b).

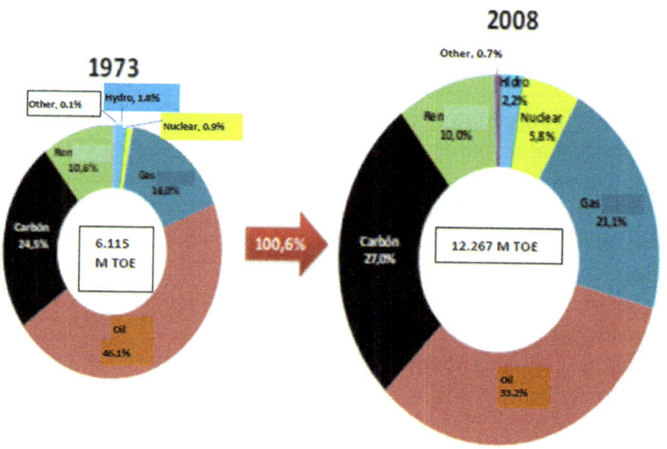

Graph 4.10 Evolution of primary energy consumption worldwide (in millions of tons of oil equivalent). *Source* Girardin (2012) based on International Energy Agency data

"oil crisis" of 1973–1974, the proportion of fossil fuels consumed globally (over total energy consumption) barely fell by 7 percentage points, from 89% in 1973 to 82% in 2008. This situation could be described as "Inertia in Energy Consumption".[71]

From this definition, it is important to point out which are some of the most relevant factors that can determine the behavior of energy consumption and in what sense they influence it. Thus, we can list some of these determinant factors:

Population Evolution: In general, it can be expected that the larger the population, the greater the energy consumption.

Evolution of Per Capita Income: Again, a positive relationship is expected between increases in this variable (measured through the gross domestic product—GDP—per capita) and increases in energy consumption.

Structure of the Economy: In this case the influence can be direct or inverse, depending on the composition of the GDP (which is the variable generally used). A productive structure based on "energy-intensive" activities surely corresponds to higher energy consumption, while a structure more linked to "knowledge-intensive" services or goods may present the same levels of "economic prosperity" with a lower incidence on energy consumption and emissions.

Technological Evolution: It will depend on what type of technology prevails. There are technologies that maximize production and economic performance in

[71] The choice of these years is not capricious: 1973 is the year of the first oil crisis of magnitude since oil (including its derivatives) became the main energy used globally, while 2008, for its part, is the year in which the "First Period of Commitment" of the PK began. Thus, 1973 was the beginning of a source substitution process that included some technological development and that allowed to improve the relative price of alternative sources. At the same time, 2008 meant, a priori, a turning point in certain production and consumption patterns in order to facilitate the transition to a less carbon-intensive economy.

the short term and others that promote energy and resource savings in general and greater efficiency in their use. In the first case, energy consumption will tend to be higher than in the second case.

Structure of the Energy Matrix: Many times the productive systems are determined by the endowment of resources (among which are included the energetic ones) and the societies in which they are inserted and the prevailing technologies in them. An energy matrix with a greater component of fossil fuels and with technologies that are not among the most efficient can tend to a greater consumption of primary energy (and, at the same time, greater emissions of GHG) than a system based on abundant hydroelectricity with greater average efficiencies for the satisfaction of the same energy needs.

Actions Against Climate Change and/or Energy Efficiency: The existence of these types of actions clearly implies a lower impact on energy consumption and also on GHG emissions.

Access to Energy, Modernization, Equity in Income Distribution, and Urbanization: These processes in general tend to increase energy consumption, mainly due to an increase in uses and greater equipment.

Pricing Policy: In some cases it may tend to increase energy consumption (low prices, subsidies) and in others to restrict it (high prices, use of fuel taxes, and energy use).

If we take into account in what sense these determinants can operate, it is easier to understand the dynamics of behavior and the evolution of energy consumption in the period of a quarter of a century pointed out. Thus, although there are many indicators that show an advance in the efficiency of resource use in general (and energy in particular), such as those that mark how GHG emissions evolved with respect to the amount of energy used, or the behavior of GHG emissions per unit of product, there are also other components that reinforce this inertia in the sense of contributing to greater energy consumption. Among the most important of the latter are, for example, the increase in population, the increase in the proportion of the urban population over the total, and the growth in per capita income.

This situation is appropriately illustrated in both Graphs 4.11 and 4.12.

Although both graphics come from different sources, they show the same phenomenon. Graph 4.11 shows the world scenario, and Graph 4.12 shows the situation of the G20 members. In the first case, the starting point is 1970, and in the other, 1990. However, in both cases there is a similar behavior. On the one hand, it can be seen how energy consumption and CO_2 emissions linked to the energy sector are increasing. However, there is a clear downward trend in two key indicators: the energy intensity of the economy (how much energy is used per unit of product) and the CO_2 emission intensity of the economy (how much is emitted per unit of product). The explanation of the behavior of both energy consumption and GHG emissions is provided by the evolution of two determining variables: the constant growth of the income per capita and the population, which in both cases drag up both energy

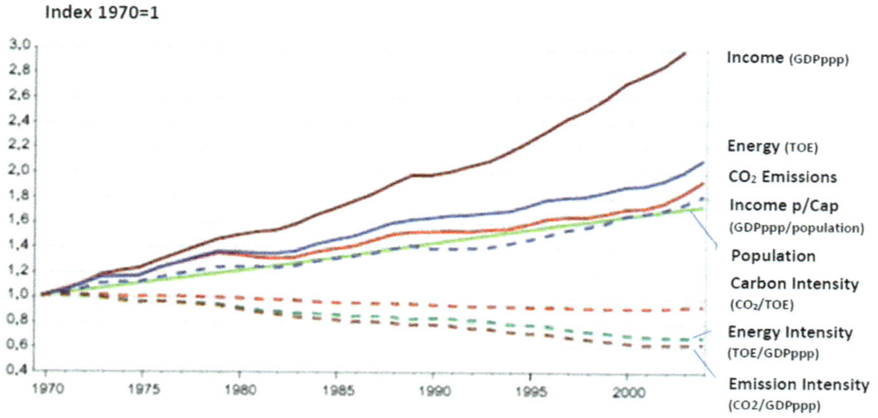

Graph 4.11 Evolution of world energy consumption indicators, global GHG emissions, average energy intensity, and some of its determinants (base 1970 = 1). *Source* Girardin (2012), taken from GEA (2012) and from data of the International Energy Agency

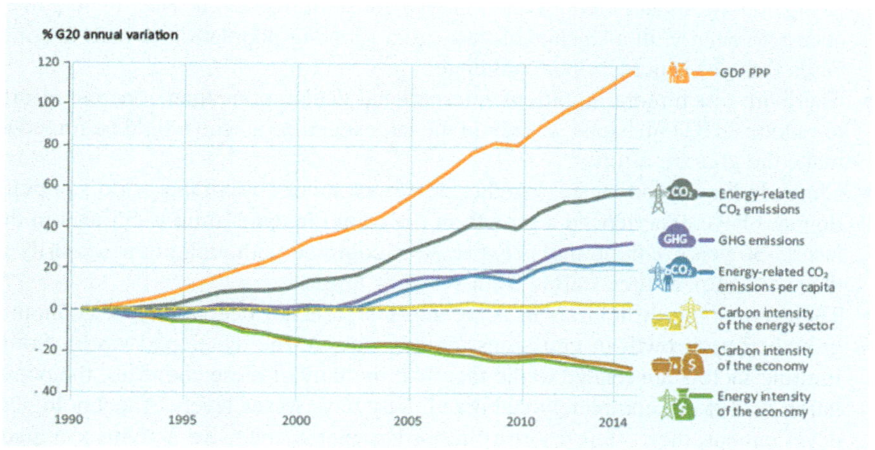

Graph 4.12 G20 countries. Evolution of energy consumption indicators, GHG emissions, energy intensity, and some of its determinants (base 1990 = 0). *Source* EGR (2017), based on Climate Transparency (2017). The G20 is a group of countries that brings together the 20 largest economies in the world and is composed of Argentina, Australia, Brazil, Canada, China, France, Germany, India, Indonesia, Italy, Japan, Mexico, Russia, Saudi Arabia, South Africa, Turkey, South Korea, the UK, and the USA

consumption and GHG emissions linked to it. This is the explanation for the presence of the aforementioned "Inertia in Energy Consumption".[72]

[72] Graph 4.11 also shows a drop in the energy consumption CO_2 emission intensity (how much is emitted per energy unit used). In Graph 4.12, the indicator is a little different, as it is composed of the

This inertia has strong implications for the effectiveness of the strategies, policies, actions, and measures to be taken with respect to climate change (and, consequently, for the prevailing uncertainty in terms of meeting the targets of the Paris Agreement). In fact, their effectiveness will depend crucially on how one can act on the determinants of energy consumption and GHG emissions linked to this consumption, taking into account the importance of GHG emissions from the sector. In this sense, the IEA shows in a report[73] that GHG emissions corresponding to electricity generation (59%) and transport (45%) increased much more than the average increase in GHG emissions between 1990 and 2007.

The uncertainties regarding the fulfillment of the PA goals (mainly linked to the possibilities of breaking this "Inertia in Energy Consumption") are also based on some empirical evidence, doubts regarding the future behavior of the relevant actors, and expectations of various kinds, such as[74]:

- By the middle of the twenty-first century, the population is expected to exceed 9 billion and the largest percentage of this population (90%) will be established in emerging and developing economies. Therefore, it is expected that a large part of this population will be composed of people who will have to improve their living conditions and increase their energy consumption levels. How to guarantee access to energy, in a sustainable way, to a growing population (which in some cases lacks it) is not a minor challenge.
- There are much more ambitious international declarations than concrete efforts to reduce GHG emissions, mainly in the largest emitters, who would be forced to make the greatest efforts.
- China, India and Indonesia, together, represent about 40% of the world's population and have experienced a growth in per capita income of up to 5 times in the last 25/30 years, a trend that is expected to continue (although not necessarily at the same rate), as they started from very low levels.
- There are some difficulties in achieving a significant decoupling of economic growth from growth in emissions, mainly outside the developed world. Some limiting factors are related to the fact that, in many of these countries, the population is still not completely stabilized, they have lower levels of technological development, they often resort to importing energy-intensive activities to guarantee sustainable levels of employment, they are focused on relatively early stages of industrialization processes based mainly on sectors linked to the exploitation of energy and natural resources, and they start from higher levels of unsatisfied basic needs. Beyond the existing difficulties in achieving the aforementioned

same energy consumption CO_2 emission intensity, but in per capita terms, and shows an increase. The presence in the G20 of China, India, and Indonesia (about 40% of the world population as a whole and starting from very low levels of energy consumption per capita compared to the world average) is those that contribute a different bias in the behavior of the variable.

[73] International Energy Agency Report (IEA/AIE) 30/05/2011. http://www.rtve.es/temas/aie-agencia-internacional-de-la-energia/48650/.

[74] Both the future scenarios and the challenges of the next 50 years are set out in detail in IEA (2011) and WBCSD (2010).

decoupling, these countries will have strong pressure on incentives to carry it out because issues linked to the mitigation of climate change will become a growing condition for non-OECD countries.

- There are significant uncertainties linked to the consequences on both energy consumption and GHG emissions of issues such as the economic growth rates presented by the various countries, the future of the nuclear industry as an alternative for electricity generation after what happened in Fukushima and the prospects of the global oil market. The existence of uncertainty does not favor technological development but generally delays it. Nevertheless, the two main challenges of the energy sector in the medium term are, without a doubt: security of oil supply and measures linked to the fulfillment of the commitments assumed in the international negotiations on climate change. In the latter case, it cannot be ignored that the policies announced in the energy sector are very far from being able to guarantee sustainable development and an increase of less than 2 °C in temperature, compared to the records prior to the Industrial Revolution, as stated in the PA.
- What will happen with the growing importance that has taken the use of coal as a fuel (which has represented a very high percentage of growth in energy consumption over the past 15 years). While coal is the fossil fuel that generates the highest emissions of GHGs (although these emissions were reduced significantly through the use of new technologies, such as fluidized bed combustion or pulverized coal), it is also the most abundant fuel and less expensive than its substitutes. In this sense, it is also important what may happen with carbon capture and storage (CCS) technologies, which could allow to extend the time horizon of fossil fuel use.
- Another issue to consider is that, in general, there is little room for maneuver to drastically change the policies that are already in place, regarding the establishment of new energy infrastructure and the replacement of the obsolete one, taking into account the long period of maturation of investments in the sector.
- Some potential scenarios to the year 2035[75] pose the following panorama:
  - A growth in global energy consumption of around 33%, in a context where a low price scenario for energy is not expected.
  - A production of 96 million barrels of oil/day, mainly of non-conventional oil (shale oil).
  - An increase in coal consumption of about 20%.
  - An increase in natural gas consumption in the order of 44%, mostly unconventional (shale gas).
  - Renewable energies and nuclear energy represent 45% of electricity generation. However, the evolution of renewable energy is expected to depend on policies that are encouraged and concentrated in relatively few countries.
  - A possible growth of tensions due to the control of non-renewable energy resources (oil and natural gas, mainly).

[75] IEA (2011).

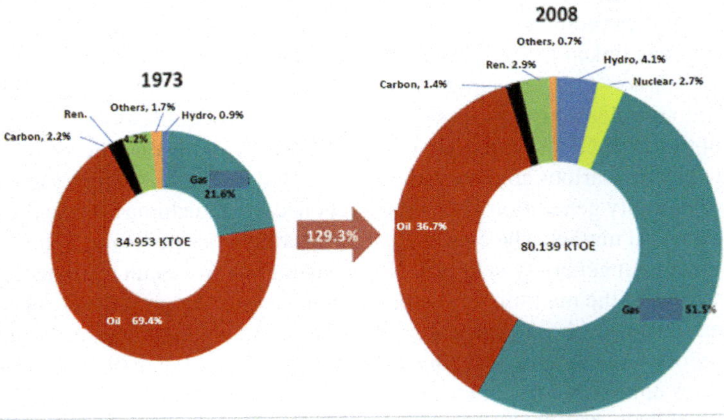

Graph 4.13 Evolution of primary energy consumption in Argentina (in millions of tons of oil equivalent). *Source* Girardin (2012), based on data from Argentina Energy Secretariat

This "Inertia" that is observed at the international level is not alien to Argentina. In fact, taking the period between 1973 and 2008, the national energy matrix did not change significantly, beyond all the efforts made to achieve that goal.

As shown in Graph 4.13, the weight of fossil fuels in the total matrix went from 93% of the total in 1973 to 90% in 2008. However, where an important change can be observed is in the composition of the fossil fuels, where natural gas has significantly replaced oil and its derivatives.

Energy consumption went from around 35 million TEPs to just over 80 million, which implies a growth of 129.3% in the 25 years analyzed. There has also been a significant growth in the proportion of hydraulic energy and the appearance of nuclear energy in the matrix.

Graph 4.14 shows the structure of electricity generation in Argentina. What can be seen from its analysis is that electricity production increased fivefold, from 21,630 to 112,382 GWH and that natural gas and hydroelectricity displaced oil and its derivatives as the main sources of electricity generation and percentages corresponding to nuclear energy and electricity imports (mainly through the Salto Grande and Yacyreta binational hydroelectric ventures) appear.

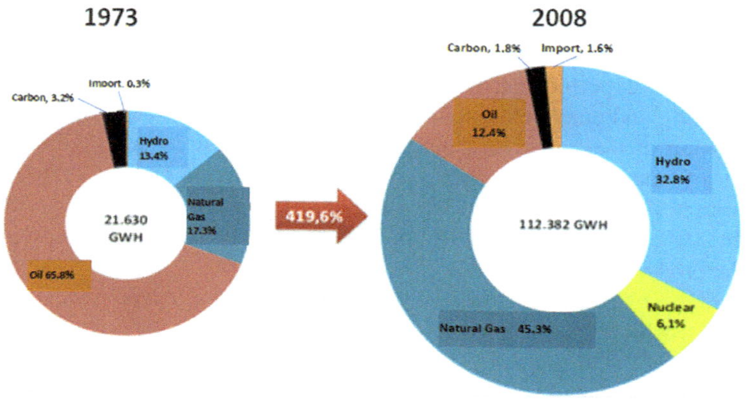

Graph 4.14 Evolution of the structure of power generation in Argentina. *Source* Girardin (2012), based on Argentina Energy Secretariat data

4.5 Challenges Related to Sustainability and the Role of the Energy Sector

The latest IPCC report[76] revived concern about global environmental issues and the challenges that humanity will have to face during the course of this century, which may become critical for the future of humanity. This issue was also addressed by other studies that raised some of the characteristics of the world that the next generations will have to confront.[77] In this framework, it is interesting to point out those issues related to the link between the energy sector and sustainability, in a context marked by an increasing degradation of the environment and the ecosystem services that such environment provides, a growing concentration of income and wealth, and demographic phenomena linked to the aging of the world population and the increase of costs related to the "care economy"[78] as a percentage of the GDP of the various countries.

Some of these challenges, which are directly linked to their potential impact on energy consumption and, consequently, on GHG emissions, are the following:

Meeting the United Nations (UN) Sustainable Development Goals (ODS) in terms of changing the development profile toward a world of low poverty, based on the assumption that there are more than 2 billion people living on an income of less than USD 2 per day; that 1.6 billion people do not have access to electricity; that

[76] IPCC (2018).

[77] IEA (2011), WBCSD (2010), UNEP (2011, 2017), and Ewing et al. (2010).

[78] The concept of the care economy refers to the system of social reproduction made up of both unpaid domestic work carried out within households, as well as the public and private provision of care services. This concept has an important gender component since, in most countries, this work is predominantly performed by women. ECLAC (2005).

900 million people do not have access to transportation; and that there are 1.8 million deaths per year due to lack of sanitation and clean water.

By 2050 it is expected that 70% of the world's population will live in cities, mainly in developing and emerging countries, which will imply an increase in the energy consumption of these people, mainly due to an increase in energy uses. To have an order of magnitude, the proportion of urban population presented the following evolution: in 1970 it represented 36%; in 2000 47%; and in 2030 it is expected to be 60%. In 2010, in China, 42% of the population was urban, and in India, for the same year, the proportion reached 29%.

The highest economic growth is expected to occur in developing or emerging economies, which are expected to account for more than 50% of global GDP by 2050.

It is estimated that changing the development profile toward a prosperous world, according to the UN goals, would mean doubling (low poverty goal) or tripling (prosperous world goal) the energy requirements by 2050.[79]

The increase in GHG emissions estimated for 2050 (52%).[80]

However, at the same time, there are also some lines of action that could involve a major change in consumption and production patterns and, thus, be able to establish a path much closer to the "decoupling" between economic growth and GHG emissions raised earlier as a response to the "Inertia in Energy Consumption".

According to UNEP (2011), if 2% of global GDP is invested annually from now until 2050 in 10 key sectors (agriculture, construction, energy, fisheries, forestry, industry, tourism, transport, water, and waste management), the result would be a low-carbon economy with more efficient use of resources but also economic growth in these same sectors. It is estimated that the transition to this "green economy" can catalyze economic activity of a size comparable to the current one but with fewer risks and impacts inherent in the current model (climate change, water scarcity, ecosystem losses) and improving the rate of employment.

According to the WBCSD (2011), the GHG emissions saved exclusively by replacing US sport utility vehicles with cars that use EU fuel economy standards are equivalent to the higher emissions due to the basic supply of electricity to 1.6 billion people who do not have access to the service.

Graph 4.15 shows the relationship between the United Nations Human Development Index and the Ecological Footprint (measured in hectares per person) of each country. Moving from left to right, we find the highest Human Development Indexes, and moving from top to bottom, the lowest impacts in terms of the Ecological Footprint. Thus, the best combinations (high Human Development Indices and low Ecological Footprint levels) would be in the lowest and rightmost quadrant of the graph.

It emerges from the analysis that most countries in Africa and some in Asia have a very low impact in terms of their pressure on resources, but at the same time, the lowest Human Development Indexes on record. Another group of countries, among

---

[79] WBCSD (2010), according to IEA (2003) data.

[80] WBCSD (2010).

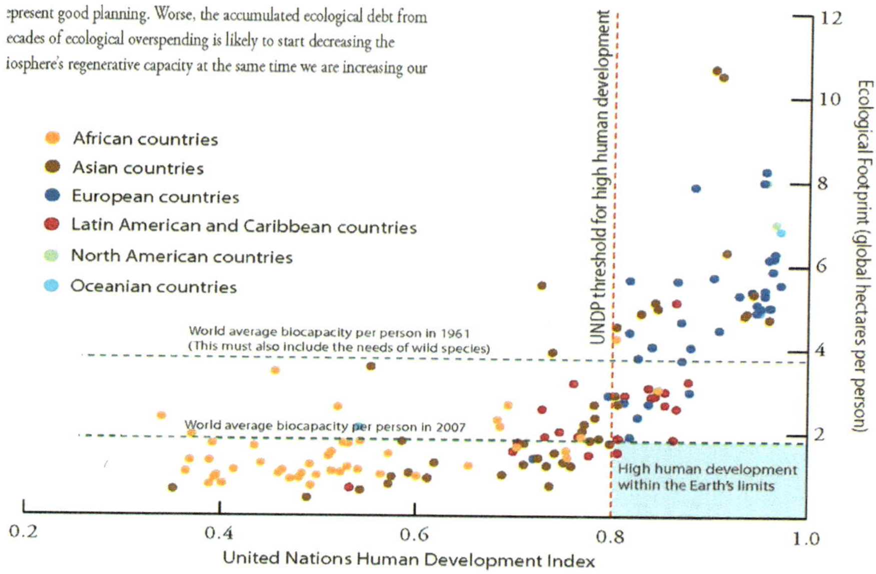

Graph 4.15 Relationship between ecological footprint (measured in hectares per person) and the United Nations Human Development Index. *Source* Ewing et al. (2010)

which are located mainly the most developed ones, have high Human Development Indices but also very high impacts on resources. Significantly, most of the countries that are in the vicinity of this "High Human Development within the Limits of the Earth" are European and Latin American and Caribbean countries. The figure also shows that these "Earth's Limits", in per capita terms, were much more "loose" in 1961 than in 2007 (the two years taken as a reference).

Graph 4.16 shows the effort of reduction and/or limitation of emissions necessary to comply with the objective of avoiding temperature increases that exceed (at the end of the twenty-first century) by 2 and 1.5 °C the values prior to the Industrial Revolution, which are set forth in the PA. This "emissions gap" states that the level of emissions compatible with the trajectory that would lead to meeting the goal of 2 °C is 42 Gt CO_2e (maximum) by 2030, and the level of emissions to meet the goal of 1.5 °C is 36 Gt CO_2e in 2030. Taking into account that current emissions exceed 50 Gt CO_2e and that they have an increasing trajectory, this implies a reduction effort of 13.5 Gt CO_2e in 2030 in the case of compliance with unconditional NDCs and 11 Gt CO_2e in the case of conditional NDCs, if the 2 °C target is to be met. If the 1.5 °C target is to be met, mitigation efforts would have to be more ambitious, reaching reductions of 19 Gt CO_2e in the case of unconditional NDCs and 16 Gt CO_2e in the case of conditional NDCs.[81]

[81] UNEP (2017).

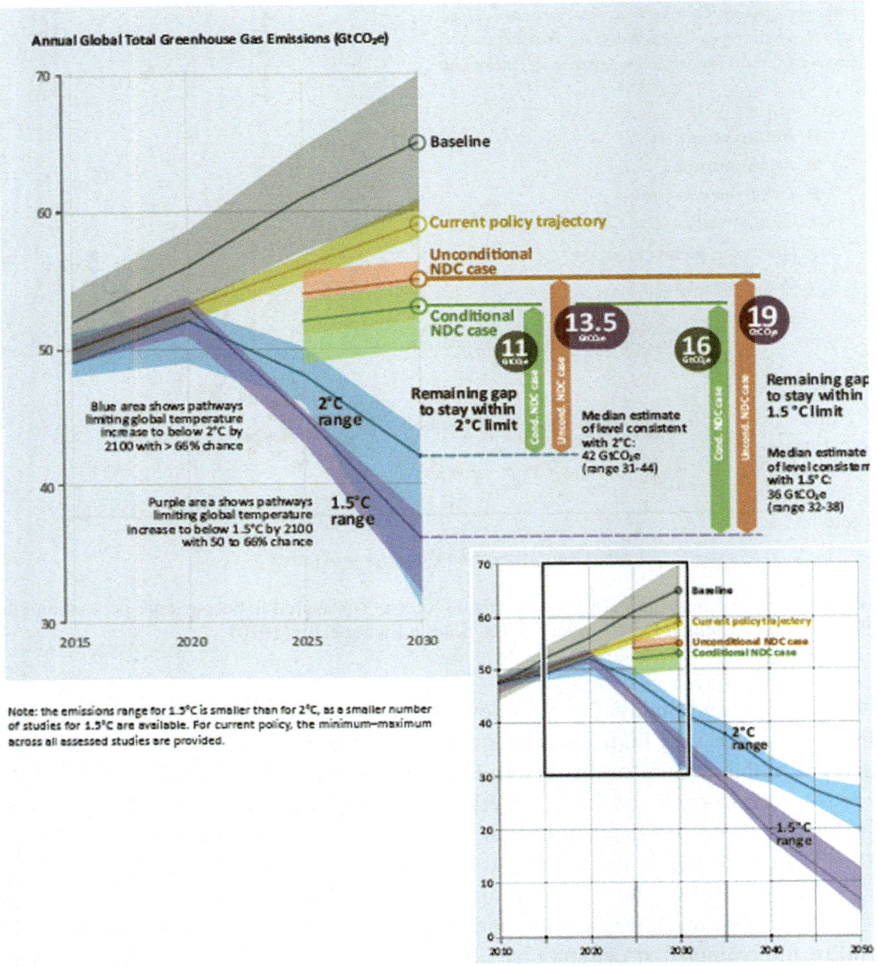

Graph 4.16 Emission gap (measured in Gt CO_2e) for the 2 and 1.5 °C targets, in the NDC compliance hypothesis (conditional and unconditional). *Source* UNEP (2017)

To achieve these reductions, it is estimated that the energy sector has one of the greatest potentials. While it is estimated that the measures that can be implemented in the agricultural sector can reach up to 6.7 Gt CO_2e, the forestry sector up to 5.3 Gt CO_2e, and other sectors (mainly waste management) 0.4 Gt CO_2e; the energy sector could contribute with 10 Gt CO_2e only in terms of electricity generation, to which could be added 4.7 Gt CO_2e in transport, 5.4 Gt CO_2e in industry, 1.9 Gt CO_2e in buildings, and 2.5 Gt CO_2e in fugitive CH_4 emissions in fuel production.

In this way, the energy sector, which is clearly part of the problem of GHG emissions linked to human activities, can also be part of the solution.

References

Aleksandrov VV, Stenchikov GL (1983) On the modeling of the climatic consequences of the nuclear war. In: Proceedings on applied mathematics, computing center. USSR Academy of Sciences, Moscow

Argawal A, Narain S (1991) Global warming in an unequal world: a case of environmental colonialism. Centre for Science and Environment, New Delhi

Argentina Republic Government (2007) Second national communication of Argentina Republic to the United Nations Framework Convention on Climate Change (2CN). Head of the Cabinet of Ministers, Secretariat of Environment and Sustainable Development, Buenos Aires, 201 pp

Argentina Republic Government (2015) Third national communication of Argentina Republic to the United Nations Framework Convention on Climate Change (3CN). Head of the Cabinet of Ministers, Secretariat of Environment and Sustainable Development, Buenos Aires, 264 pp

Argentina Republic Government (2016) First review of the national determined contribution (NDC) (17112016). Available in http://www4.unfccc.int/ndcregistry/PublishedDocuments/Argentina%20First/17112016%20NDC%20Revisada%202016.pdf

Argentina Republic Government (2017a) Second biennial update report of Argentina Republic to the United Nations Framework Convention on Climate Change (BUR2). Ministry of Environment & Sustainable Development of Argentina, Buenos Aires, 86 pp

Argentina Republic Government (2017b) Second biennial update report of Argentina Republic to the United Nations Framework Convention on Climate Change (BUR2) Annex 1 & 2. Buenos Aires, 171 pp

Azqueta Oyarzún D (1994) Economic valuation of environmental quality. McGraw-Hill, Madrid

Barros V, Menéndez A, Nagy G (2005) Climate change in the Río de la Plata. CIMA/CONICET. Selection of technical reports of the projects "Impacts of global change in the coastal areas of the Río de la Plata" and hydroclimatic vulnerability of the Río de la Plata estuary: human influence, ENSO and trophic state. Project "Assessment of impacts and adaptations to climate change (AIACC)", May 2005. START-TWAS-UNEP, Buenos Aires. Available in http://www.cima.fcen.uba.ar/~lcr/libros/Cambio_Climatico-Texto.pdf

Baumert K, Herzog T, Pershing J (2005) Navigating the numbers. Greenhouse gas data and international climate policy. World Resources Institute (WRI), Washington, DC, 132 pp

Bhaskar V (1995) Distributive justice and the control of global warming. In: Bhaskar V, Glyn A (eds) The north, the south and the environment. Ecological constraints and the global economy. UNU, Tokyo

Climate Transparency (2017) Brown to green. The G20 transition to a low-carbon economy. Climate Transparency, c/o Humboldt-Viadrina Governance Platform, Berlin, Germany. www.climate-transparency.org

Crutzen PJ, Birks JW (1982) Twilight at noon: the atmosphere after a nuclear war. Ambio 11:114

Ewing B, Moore D, Goldfinger S, Ousler A, Reed A, Wackernagel M (2010) Ecological footprint atlas 2010. Global Footprint Network, Oakland

Food and Agriculture Organization—FAO (2008) Forest & energy. Key issues. United Nations Food and Agriculture Organization (FAO), 86 pp. Available in http://www.fao.org/docrep/pdf/010/i0139s/i0139s00.pdf

Funtowicz S, Ravetz J (1993) Political epistemology. Science with people. In: Fundamentals of human sciences series, No. 107. CEAL, Buenos Aires

Girardin LO (2000) Global climate change and the distribution of mitigation costs of its eventual consequences among different countries. Buenos Aires. Available in http://fundacionbariloche.org.ar/

Girardin LO (2012) Energy & climate change: the challenges of low carbon development. In: Conference at the Buenos Aires headquarters of the Fundación Patagonia Third Millennium, Buenos Aires, Aug 9. http://www.patagonia3mil.com.ar/

Girardin LO (2013) Socioeconomic and political aspects of climate change. In: From the convention to the Kyoto Protocol (1990–2000). Fundación Patagonia Third Millennium, Trelew-Buenos

Aires. ISBN 978-987-26155-8-1. Available in http://www.patagonia3mil.com.ar/wp-content/uploads/libros/publicaciones-politicas_ambientales.pdf
Girardin LO (2017a) The economics of climate change in Argentina. Volume I: synthesis report and report on economic valuation. United Nations Economic Commission for Latin America and the Caribbean (ECLAC), National Council for Scientific and Technological Research of Argentina (CONICET), Fundación Patagonia Third Millennium (FUNPAT), Buenos Aires, Trelew, 366 pp. ISBN 978-987-45525-3-2. Available in http://www.patagonia3mil.com.ar/publicaciones/
Girardin LO (2017b) The economics of climate change in Argentina. Volume II: impacts, vulnerability & adaptation. United Nations Economic Commission for Latin America and the Caribbean (ECLAC), National Council for Scientific and Technological Research of Argentina (CONICET), Fundación Patagonia Third Millennium (FUNPAT), Buenos Aires, Trelew, 490 pp. ISBN 978-987-45525-4-9. Available in http://www.patagonia3mil.com.ar/publicaciones/
Girardin LO (2017c) The economics of climate change in Argentina. Volume III: emission scenarios & mitigation. United Nations Economic Commission for Latin America and the Caribbean (ECLAC), National Council for Scientific and Technological Research of Argentina (CONICET), Fundación Patagonia Third Millennium (FUNPAT), Buenos Aires, Trelew, 564 pp and statistics Annex. ISBN 978-987-45525-5-6. Available in http://www.patagonia3mil.com.ar/publicaciones/
Global Energy Assessment—GEA (2012) Global energy assessment. Towards a sustainable future. In: Johansson TB, Patwardhan A, Nakicenovic N, Gómez-Etcheverri L (eds) International institute for applied system analysis (IIASA). Cambridge University Press, New York
Herzer H (1990) Disasters are not as natural as they seem. Environ Urban 30:3–10. IIED-AL, Buenos Aires
Intergovernmental Panel on Climate Change—IPCC (1990) Climate change. The IPCC scientific assessment. In: Report prepared for IPCC by Working Group I. Cambridge University Press, Cambridge. Available in www.ipcc.ch
Intergovernmental Panel on Climate Change—IPCC (1992) Climate change. The 1990 and 1992 IPCC assessments. In: IPCC first assessment report overview and policy makers summaries and 1992 IPCC supplement, 180 pp
Intergovernmental Panel on Climate Change—IPCC (1994) Climate change 1994. In: Radiative forcing of climate change and an evaluation of the IPCC IS92 emission scenarios. Cambridge University Press, New York, 345 pp
Intergovernmental Panel on Climate Change—IPCC (1995) Climate change 1995. The science of climate change. In: Contribution of Working Group I to the second assessment report of the IPCC. Cambridge University Press, Cambridge, 588 pp. Available in www.ipcc.ch
Intergovernmental Panel on Climate Change—IPCC (2007a) Climate change 20007. The physical science basis. In: Working Group I contribution to the IPCC fourth assessment report (4AR). Cambridge University Press. Available in www.ipcc.ch
Intergovernmental Panel on Climate Change—IPCC (2007b) Climate change 2007. Impacts, adaptation and vulnerability. In: Working Group II Contribution to the IPCC fourth assessment report (4AR). Cambridge University Press. Available in www.ipcc.ch
Intergovernmental Panel on Climate Change—IPCC (2007c) Climate change 2007. Mitigation and climate change. In: Working Group III contribution to the IPCC fourth assessment report (4AR). Cambridge University Press. Available in www.ipcc.ch
Intergovernmental Panel on Climate Change—IPCC (2013) Climate change 2013. The physical science basis. In: Working Group I contribution to the IPCC 5AR. Cambridge University Press. Available in www.ipcc.ch
Intergovernmental Panel on Climate Change—IPCC (2014a) Climate change 2014. Impacts, adaptation and vulnerability. In: Working Group II contribution to the IPCC 5AR. Cambridge University Press. Available in www.ipcc.ch
Intergovernmental Panel on Climate Change—IPCC (2014b) Climate change 2014. Mitigation and climate change. In: Working Group III contribution to the IPCC 5AR. Cambridge University Press. Available in www.ipcc.ch

Intergovernmental Panel on Climate Change—IPCC (2018) Global warming of 1.5 °C. An IPCC special report on the impacts of global warming of 1.5 °C above pre-industrial levels and related global greenhouse gas emission pathways, in the context of strengthening the global response to the threat of climate change, sustainable development, and efforts to eradicate poverty. Available in http://www.ipcc.ch/report/sr15/

International Energy Agency—IEA/AIE (2003) CO_2 emissions from fuel combustion 1971–2001

International Energy Agency—IEA/AIE (2007) World energy outlook 2007. Paris

International Energy Agency—IEA/AIE (2011) World energy outlook 2011. Paris, 666 pp

Lipietz A (1995) Enclosing the global commons: global environmental negotiations in a north-south conflictual approach. In: Bhaskar V, Glyn A (eds) The north, the south and the environment. Ecological constraints and the global economy. UNU, Tokyo

Martínez Alier J (1995) Distance learning course on ecological economics. Environmental training network. United Nations Environment Program—UNEP, Quito

Meyer A (2000) Contraction & convergence—the global solution to climate change. Green Books, Devon, 96 pp

National University of Litoral (Argentina)-UNL (2003) Technical report on Santa Fe floodings. Tuesday 3 of June 2003. Available in https://www.unl.edu.ar/noticias/news/view/informe_t%C3%A9cnico_sobre_la_inundaci%C3%B3n_de_santa_fe#.W88kQVRKjIU

Nordhaus WD (2007) The challenge of global warming: economic models and environmental policy. Yale University, New Haven, CT, 200 pp. Available in http://aida.econ.yale.edu/~nordhaus/homepage/dice_mss_091107_public.pdf

Olivier JGJ, Janssens-Maenhout G, Muntean M, Peters JAHW (2016) Trends in global CO_2 emissions: 2016 report, Nov 2016. European Commission, Joint Research Centre (JRC), Directorate C—Energy, Transport and Climate, PBL Netherlands Environmental Assessment Agency, The Hague. JRC103425, PBL2315. Internet: http://edgar.jrc.ec.europa.eu/news_docs/jrc-2016-trends-in-global-co2-emissions-2016-report-103425.pdf

Postigo JC et al (ed) (2013) Climate change, social movements and public policies: a necessary linkage. In: Working Group: climate change, social movements and public policies. Latin American Council of Social Sciences (CLACSO), Alejandro Lipschutz Institute of Sciences (ICAL), Santiago de Chile

Stern N (2006) The Stern review: the economics of climate change. Cambridge University Press. www.sternreview.org.uk

Turco RP, Toon OB, Ackerman TP, Pollack JB, Sagan C (1983) Nuclear winter: global consequences of multiple nuclear explosions. Science 222(4630):1283–1292. https://doi.org/10.1126/science.222.4630.1283

United Nations Economic Commission for Latin America and the Caribbean—ECLAC (2005) Care economics and economic policy. An approach to their interrelationships. Corina Rodríguez Enriquez. In: 38th meeting of the board of directors of the regional women's conference of Latin America and the Caribbean. Panel "social protection policies, care economy and gender equality", 7 and 8 Sept 2005. ECLAC, Mar del Plata

United Nations Economic Commission for Latin America and the Caribbean—ECLAC (2012) The economics of climate change in Chile. Santiago de Chile, 134 pp

United Nations Economic Commission for Latin America and the Caribbean—ECLAC (2014) The economics of climate change in Argentina. First approach. Santiago de Chile, 241 pp

United Nations Environmental Programme—UNEP (2011) Towards a green economy. Guide for sustainable development and the eradication of poverty. Synthesis for policymakers. Nairobi. www.unep.org/greeneconomy

United Nations Environmental Programme—UNEP (2017) The emission gap report 2017. A UN environment synthesis report. Nairobi, Nov 2017. https://wedocs.unep.org/bitstream/handle/20.500.11822/22070/EGR_2017.pdf

United Nations Framework Convention on Climate Change—UNFCCC (1992) United Nations Framework Convention on Climate Change. Available in Spanish in https://unfccc.int/resource/docs/convkp/convsp.pdf

United Nations Framework Convention on Climate Change—UNFCCC (1997) Kyoto Protocol. Available in Spanish in https://unfccc.int/resource/docs/convkp/kpspan.pdf
United Nations Framework Convention on Climate Change—UNFCCC (2010) FCCC/CP/2010/7/Add.1. https://unfccc.int/sites/default/files/resources/docs/2010/cop16/spa/07a01s.pdf
United Nations Framework Convention on Climate Change—UNFCCC (2012) Doha Amendment to the Kyoto Protocol. Available in Spanish in https://unfccc.int/files/kyoto_protocol/application/pdf/kp_doha_amendment_spanish.pdf
United Nations Framework Convention on Climate Change—UNFCCC (2015) Paris Agreement. Available in Spanish in https://unfccc.int/files/meetings/paris_nov_2015/application/pdf/paris_agreement_spanish_.pdf
United States Energy Information Administration—USEIA (2007) International energy outlook 2007. Washington, DC. Available in www.eia.doe.gov/oiaf/ieo/index.html
World Business Council on Sustainable Development—WBCSD (2010) Vision 2050. The new agenda for business. Available in https://www.wbcsd.org/Overview/About-us/Vision2050
World Business Council on Sustainable Development—WBCSD (2011) Collaboration, innovation, transformation. Ideas and inspiration to accelerate sustainable growth—a value chain approach. Available in https://www.wbcsd.org

Chapter 5
Socioeconomic and Political Aspects of Climate Change. The Role of the Clean Development Mechanism and Other Market-Based Mechanisms in Contributing to the Ultimate Objective of the UNFCCC and Sustainable Development. A Latin American Point of View of the Situation After the Paris Agreement

Abstract This chapter analyzes the role of the Clean Development Mechanism and other Market-based Mechanisms in contributing to the ultimate objective of the United Nations Framework Convention on Climate Change and Sustainable Development. A Latin American Point of View of the situation after the Paris Agreement. Latin America was a pioneer region in terms of its early participation in emission reduction and/or limitation mechanisms. This situation occurred even before the issue was installed with force in other regions, later very active in the use of the mechanism, mainly China and Southeast Asia. However, this did not result in "advantages" for the region, in terms of investment settlement or to establish better conditions in the international negotiation within the framework of the UNFCCC. This situation leads to what in some previous documents has been called "certain discouragement of the region with respect to the CDM". It is interesting to explore how this can play in the deepening of the commitments of Latin America with the prevention of climate change, and although this objective far exceeds the scope of this brief document, we will try to raise some issues that cannot be overlooked and try to interpret the problem. More than 20 years after the Kyoto Protocol (KP), a consensual, binding, and ambitious international climate agreement is still being sought. However, substantive decisions continue to be postponed. In short, it is impossible for a market to survive without demand. And the demand arises from the degree of rigor that exists in the fulfillment of the commitments that are assumed and to what extent these commitments imply a real effort to reduce emissions.

Keywords Market instruments · Clean Development Mechanism · Kyoto Protocol · Paris Agreement · Climate change · Negotiations

---

Leonidas Osvaldo Girardin. Argentina's National Council on Science and Technology (CONICET) and Fundacion Bariloche (FB). December 2019. This is an updated version of the article by Girardin (2018).

L. O. Girardin, *Socioeconomic and Geopolitical Aspects of Global Climate Change*, The Latin American Studies Book Series, https://doi.org/10.1007/978-3-031-53246-7_5

5.1 Introduction

The Paris Agreement (PA)[1] is an international legal instrument that emerged within the United Nations Framework Convention on Climate Change (UNFCCC) that establishes measures for the limitation and/or reduction of greenhouse gas (GHG) emissions. It was signed on December 12, 2015, in the context of the 21st Conference of the Parties to the UNFCCC (COP-21) and entered into force on November 4, 2016, when the requirements for it were met. The PA covers the period after 2020 and has as its main long-term objective to keep the increase in global average temperature below 2 °C with respect to pre-industrial levels and, additionally, to commit governments to continue working to limit it to 1.5 °C.

In this context, the "Parties" to the UNFCCC presented general national action plans, called Nationally Determined Contributions (NDCs), through which they implement the measures and actions that are expected to limit and/or reduce their emissions. In addition, they agreed to report their contributions every five years (in order to set more ambitious targets); they agreed to inform each other and to tell the society of the extent to which their targets have been met (to ensure transparency and monitoring). Likewise, those Parties that integrate the group or the most developed countries committed to financing the fight against climate change (to help developing countries both reduce their emissions and increase resilience to the effects of climate change).

Article 6 of the PA provides for an approach that promotes voluntary cooperation between "Parties" for the enforcement of NDCs and the promotion of sustainable development. It also provides for the generation of International Transfer Mitigation Outcomes (ITMOs) between Parties, which may include Emission Reduction Units (ERUs). It is impossible to ignore the similarities and relationship between this Article 6 of the PA and the Clean Development Mechanism (CDM) arising from Article 12 of the Kyoto Protocol (KP).[2]

Given this situation, it is good to analyze what happened to the expectations that were originally held about the role that both the CDM and other Market Mechanisms could play in the fight against climate change, particularly for Latin America and the Caribbean (LAC) and for the Argentine Republic in particular.

---

[1] UNFCCC (2015).

[2] The CDM was one of the "Flexibility Mechanisms" that emerged from the KP to facilitate compliance with the Quantified Emission Reduction and/or Limitation Commitments of Annex I Parties to the UNFCCC. Among the different mechanisms created, the main ones, besides the CDM, were Joint Implementation (Article 6) and Emissions Trading (Article 17).

5.2 Latin America: From Euphoria to Disenchantment

Latin America was a pioneer region in terms of its early participation in the mechanisms for reducing and/or limiting GHG emissions through not only what was the Pilot Phase of Joint Implementation (JI) projects (which began operating even before the existence of the Kyoto Protocol), but also the early stages of the CDM, once the KP was established. This situation occurred even before the issue was strongly installed in other regions, which later became very active in the use of the mechanism, mainly China and South East Asia.[3] Not only that: many of the government offices in the Latin American Region[4] have shown themselves, since their beginnings, to be among the most dynamic in the process of promoting project activities that can be applied to these mechanisms, based on the identification of mitigation options in relevant sectors for this purpose and the attempt to attract investment opportunities in these projects.

However, paradoxically, more than 20 years after the signing (December 11, 1997), more than a decade after the KP came into force (February 16, 2005), and a long time since the appearance of the Modalities and Procedures (M&P) that regulate CDM project activities (November 2001), it is clear that two very marked phenomena have occurred. On the one hand, the most successful experiences in terms of quantity and diversification of CDM projects are presented in countries whose private sector is more dynamic in participating in this type of mechanism, regardless of whether or not the state has explicit policies to support these initiatives. In this sense, the paradigmatic case in Latin America is Brazil, which does not and did not have a CDM Promotion Office, but only a Designated National Authority (DNA) and yet was always at the forefront in the application of this mechanism in LAC. On the other hand, the region has suffered and continues to suffer from what could be called the "perverse logic of the CDM",[5] because many of the lower cost mitigation measures or those with a higher volume of emission limitation and/or reduction (mainly Fuel Substitution in Electricity Generation, in the first case, and the development of Non-GHG-emitting Energies, in the second case) have been carried out, mainly between the 70s and the 90s, thus already forming part of their baselines and making project activities (which can be considered additional[6]) relatively "more expensive", mainly when compared to the opportunities available to regions which have delayed the

[3] See Figueres (2002) and CAEMA (2003).

[4] In this document, when we refer to government offices, we are referring not only to the Designated National Authorities (DNA) but also to the Promotion Offices that in many countries developed in parallel with those mentioned above.

[5] See Girardin (2000, 2013) and Gobierno de la República Argentina (1999).

[6] "Additionality" in this context relates to the fact that the activity is only feasible to be developed because of the existence of the CDM; otherwise it would not be carried out for various reasons (barriers), which the presence of the mechanism helps to overcome.

implementation of such measures over time (such as South East Asia, for example) and which are competing to impose their projects in the same field.[7]

This situation leads to what some previous documents have called "a certain discouragement of the region with regard to the CDM".[8] It is interesting to explore how this may play into the deepening of Latin America's commitments to the prevention of climate change, and although this objective far exceeds the scope of this brief document, an attempt will be made to raise some issues that cannot be overlooked when trying to interpret the problem. However, it is also important to note that some of these issues have been emerging for a long time and, far from being resolved, in some cases they have been deepened.[9]

5.3 Climate Change and Heterogeneity

Although climate change is a global phenomenon, empirical evidence indicates that the geographical distribution of the effects will be very heterogeneous, which makes it even more difficult to plan appropriate policies to overcome them. This heterogeneity in the expected impacts of climate change will overlap on the heterogeneities and inequalities already existing in other fields, among different countries, regions, sectors, activities, and social groups, so everything indicates that the incidence of climate change will be different on all of them, depending on their degree of vulnerability. To make this panorama even more complex, all the regional studies on the expected impacts of climate change indicate that the consequences that the poorest countries (and within them the most unprotected social groups) will have to bear are disproportionately greater than their very limited responsibility in having reached the current situation.[10]

The degree of vulnerability of different countries, regions, socioeconomic sectors, activities, and communities to these phenomena is closely related to their capacity to absorb, cushion, or adapt to the effects of these changes.[11] This situation, in turn, will depend on the possibility of having technologies, infrastructure, and means suitable for this purpose, and in this sense, the poorest populations, the most climate-dependent activities, and the countries and/or regions with less diversified economic structures will most likely present greater degrees of vulnerability. This situation may lead to a widening of the North–South gap, but also to a deepening of inequalities within countries themselves, regardless of the historical responsibilities of each of

---

[7] See, among others, Bouille et al. (1999, 2000), Girardin and Di Sbroiavacca (2000), Girardin and Bouille (2002, 2003), and Girardin (2008a).

[8] Girardin and Bouille (2002, 2003) and Girardin (2008a), among others.

[9] See, for example, Gobierno de la Republica Argentina (1999) and Girardin (2000).

[10] IPCC (1998, 2014), Girardin (2008b), and UNFCCC (2006, 2007).

[11] IPCC (2014), Herzer (1990), and Girardin (2000). After Herzer, there were authors who deepened the issue of social vulnerability to climate change. See, among others, for the case of Argentina, mainly Natenzon et al. (2006).

the actors involved, in terms of their contribution to the climate change. In reality, we are referring to a North–South conflict that is more sociological than geographical in nature,[12] bearing in mind that phenomena such as Hurricane Katrina, which hit New Orleans in 2005, demonstrated that even the most powerful country of the North contains its own South and that in our South, the dominant elites have patterns of consumption of energy, goods, and natural resources that are equal to or superior to those of many of the richer groups in the North.[13]

The presence of degrees of uncertainty and heterogeneity such as those mentioned will influence decision-making, given that decisions should be taken despite the lack of certainty about the true level of knowledge about the future consequences to be faced. However, all the estimates made about the potential repercussions that climate change may bring are of such a magnitude that they justify some kind of intervention to avoid them, applying strategies, policies, and preventive measures based on the precautionary principle.[14]

In this sense, the possibilities of taking action more immediately to mitigate the potential effects of climate change focus on those causes related to human activities that are known to influence the net amounts (gross emissions minus removals) that are emitted of GHGs, leading to the need to reduce, limit, and/or avoid these emissions in key sectors such as energy, industry, agriculture and livestock, waste management, land use, land use changes, and forestry.

This is where a significant amount of hope was placed that the CDM could contribute to the fact that, in non-Annex I countries (NAI),[15] the consumption and production patterns that accompanied higher levels of development were not necessarily those followed by industrialized countries (IPs) to reach their current level of economic development.

5.4 Resource Allocation and Climate Change (I): Who Pays and Departing from Which Argument?

An important issue to consider is that each strategy, policy, and/or concrete measure adopted to limit GHG emissions implies certain types of impacts on the activities involved and, consequently, certain types of sacrifices on the economies of the societies that implement them. It is not casual that one of the most conflictive points of negotiation on the international climate change agenda is related to the distribution of mitigation costs among the different countries. The problems that each society has to face are different, as are the degrees of vulnerability to which they are exposed.

---

[12] See Girardin (2000). Definition taken from Lipietz (1995).

[13] Girardin (2013).

[14] This principle states that, when the future effect of a present cause is uncertain, but may be very harmful and irreversible, it is wiser to act immediately to remove the best known cause from those upon which action can be taken. See Girardin (2000, 2013).

[15] UNFCCC (1992).

Even the interests of the various actors can be conflicting depending on the modality adopted to address climate change, which will necessarily lead to the application of different approaches to the issue.

From the economic point of view, the solution that is finally adopted will not be neutral in terms of the effects on the distribution of income among the various countries, regions, sectors, activities, and social groups linked to them. Different methodological approaches will determine different results, depending on the models and assumptions used to formulate and simulate possible future scenarios. In some cases, this relationship between the assumptions and logical structure of the models used and the results obtained is so close that this situation adds even more uncertainty to the true mitigation costs that each actor involved will have to face.[16] Nevertheless, there is consensus that the first emissions limitations will be less costly per ton avoided/reduced, while the opportunities for lower costs will be taken advantage of at the outset, and that these costs will gradually increase as these opportunities are exhausted and actions have to be taken in sectors that present less advantageous options. Thus, a major point of contention is which strategy each country will choose and how the burden of GHG mitigation costs will be shared among the different countries.

As results evident, the predominant position among these countries that have already made emission reduction commitments under the UNFCCC and the KP is based on prioritizing economic efficiency at global level over all other criteria, advocating that emission reductions be made where it is cheapest to obtain them using cost-effectiveness approaches. Acting in this way dilutes their greater historical responsibility for having reached this situation and transfers a large part of the mitigation measures carried out to the less developed countries, whose natural resources, wages, and other factors of production and other fundamental elements in this equation are cheaper.[17] For their part, the rest of the countries (those that in the UNFCCC and the KP had not assumed quantified commitments to limit and/or reduce emissions by virtue of the recognition of their lesser historical responsibility in the process of climate change) seek to make this historical responsibility a determining criterion when it comes to sharing the burden of facing the expected impacts of climate change, while they argue that the objective of minimizing the costs of mitigating GHG emissions should not hide the difference in responsibility that exists between countries (a fact that is recognized in the UNFCCC itself), which refers to "common but differentiated responsibilities" and that developed countries should show initiative in preventing the impacts of climate change, nor to fail to take into consideration ethical values such as that all inhabitants of the earth have the same right to enjoy the benefits of economic development and thus access to adequate

---

[16] While there is a general consensus that emission reductions are lower in NAI countries than in IPs, this is not necessarily always true. Some studies developed by the Institut d'Économie de et de Politique de l'Énergie (IEPE), of the University of Grenoble, show that at the regional level the opposite is often true, as the cost of GHG emission limitations depends more on the situation from which they are based (baseline) than on the relative level of development of the area in which the measure is applied. See Criqui and Kouvariatkis (1997), cited in Bouille et al. (1999).

[17] See previous footnote.

levels of comfort and quality of life. In this process of fighting poverty and seeking greater well-being for their excluded population, many of these countries are likely to increase their current GHG emissions.

However, obviously, not everything is so linear. Not only do emissions matter in "relative" terms, but also "absolute" emissions, because it is these that will contribute to the increase in atmospheric concentrations of GHGs over time and, consequently, to changes in temperatures and other climate variables and parameters. In this sense, the growing importance of emissions from countries such as China, India, or Indonesia cannot be ignored, just to take a few examples.

Thus, from an economic point of view there are two fundamental issues related to climate change: not only who should pay, but also what use should be prioritized to allocate the limited funds that are available, in relation to the number of issues to be addressed. As resources that are dedicated to certain policies and measures will not be available for alternative uses, less developed countries will have to decide between allocating resources for adaptation or allocating them for mitigation.

5.5 Resource Allocation and Climate Change (II): Synergies and Conflicts Between Adaptation and Mitigation

The most recent data provided by experts on the subject, contained in both the Fourth Assessment Report (4AR) and the Fifth Assessment Report (5AR) produced by the Intergovernmental Panel on Climate Change (IPCC) in 2007 and 2013–2014, respectively,[18] leave no doubt about the growing influence of human activities on this process. But neither do they leave any doubt that, beyond the efforts to mitigate GHG emissions made by countries such as, for example, Argentina (which emits less than 1% of total world emissions),[19] these countries will be obliged to make a major effort to adapt to the impacts of climate change, which they will inevitably suffer.

Developing countries (DCs) are more vulnerable to the potential impacts of climate change even though the historical responsibility (and also current responsibility in the case of the vast majority of them) in the process that led to the current situation is less.[20] But, additionally, given the high concentration of GHG emissions in a few countries and the low share of current emissions in most DCs, the effects of limiting GHG emissions in most of those DCs would not have significant consequences for solving the problem of preventing increases in atmospheric GHG concentrations, as discussed above. Even if they implement mitigation policies and fully comply with the commitments and plans to carry them out, it will still be

---

[18] IPCC (2007, 2014). At the time of writing the present article, the Sixth Assessment Report (6AR) is being prepared and is expected to be completed in 2022.

[19] 0.89% of the world total, according to the WRI (2016). http://www.wri.org/blog/2016/04/when-could-paris-agreement-take-effect-interactive-map-sheds-light-es.

[20] IPCC (1998) and UNFCCC (2007).

necessary to carry out some degree of adaptation to the expected impacts of climate change, which they will suffer anyway.[21]

Thus, they will also face significant adaptation costs. However, the paradox is that most of the funds available at the international level to address climate change issues are allocated to activities related to mitigation (the main responsibility of the most developed countries) rather than to adaptation (the main urgency of DCs and mainly least developed countries—LDCs), which constitutes an additional barrier for the most vulnerable countries to meet the challenges of climate change.

One of the main arguments of developed countries to justify the lack of funding for climate change adaptation activities in developing countries starts from considering adaptation as a local, or at least, a national, issue, instead of considering it as a global problem, as they do with mitigation. If this is the case within the Global Environment Facility (GEF), no significant amount of funding will be dedicated to adaptation, as it mainly finances the incremental costs incurred in addressing a global problem.[22]

However, adaptation must necessarily be considered a global problem from at least two points of view: (a) firstly, because developing countries are forced to adapt to climate change and variability regardless of their responsibility for the origin of the problem and, (b) secondly, because without joint responsible action by all the actors involved, it will not be possible to adapt to the changes.[23] In any case, adaptation is a global problem that has different ways of being addressed at both the national and local levels, depending on the national circumstances of each country. These national circumstances have a fundamental influence on two aspects: (a) the degree of incidence of the potential impacts of climate change and (b) the response capacity of each society.

If we take into account that those most vulnerable to the expected impacts of climate change are also generally the most vulnerable to all kinds of changes in the starting conditions (changes in the process of business globalization, changes in the prices of raw materials and energy prices, etc.), it is not unreasonable to think about the implementation of development policies as the best way to start developing strategies to adapt to climate change. A fairer, more egalitarian, better educated, and informed society, with better levels of health, is much better prepared to face all the challenges, not only those related to climate change.

---

[21] IPCC (2014).

[22] Incremental costs are those incurred in carrying out activities dedicated to generating global benefits that are additional to the costs incurred in actions aimed at obtaining local benefits. This recognizes the "incremental" effort made to address a global problem.

[23] Moreover, in an international context where constant reference is made to the process of "globalization" of business, the movement of capital, and the transfer of information, it sounds a little hypocritical to pretend that adaptation to the impacts of climate change is seen as an exclusively local problem.

5.6 The Role of the Market and the CDM in Contributing to Sustainable Development: "From Saying to Doing …"[24]

Estrada Oyuela (2008), more than 10 years ago already raised two questions that are impossible to evade if anybody wants to understand what is happening today with the generically called "carbon markets": (a) the purpose of the KP was to reduce and limit GHG emissions to stabilize their atmospheric concentrations, as agreed in the UNFCCC, not the creation of a carbon market; and (b) in addition the reduction of emissions required in the period 2008–2012 to the KP Parties has been significantly lower than the availability of credits.[25]

In this regard, according to calculations by the Japanese Ministry of Economy, Trade and Industry, cited by Estrada Oyuela (2008), the potential supply of carbon credits expected for the first commitment period (2008–2012) was estimated at 10.6 billion tons of CO_2e (4.4 billion for the Russian Federation, 2.4 billion for Ukraine, 1.500 million for the 12 new members of the European Union (EU), and finally 2.3 billion for certified emission reductions or CERs under the CDM), while potential demand for the same period was estimated at only 2.114 billion tons (1.5 billion from the EU, 200 million from New Zealand, Switzerland, Norway, and other Annex I countries, 260 million from Japan, and 34 million from Australia). Under these conditions, it was already clear that little could be expected from the CDM to redistribute income between the rich and the poor. There was speculation about how the potential entry of the USA into an analogous system might play (probably by increasing transaction prices through a strong increase in demand for emission reductions), but this never happened to the extent that was expected to energize the system.

Additionally, to get an idea of what the situation was like, at the end of the KP's first commitment period (2008–2012), according to the information provided on the UNFCCC Secretariat's website as of December 31, 2012,[26] the CDM also had a limited scope in its role in the technology transfer process and in its contribution to the sustainable development of the countries hosting the project activities.

At the time of the end of the KP's first commitment period (2008–2012), there were more than 7500 projects in the pipeline at the CDM Executive Board (CDM-EB). While there was no precise information on the average annual number of CERs they generated, it was estimated that by the end of 2012 they would have generated more than 2.216 million tons and by 2015 they were expected to be generating a potential supply of more than 4.76 billion CERs. At that time, 5511 projects were registered that would have generated and over 2191 million CERs by the end of 2012, while

[24] There is a popular saying in Latin America and Spain: "from saying to doing, there's a long way", it means that things are easier said than done.

[25] Estrada Oyuela (2008).

[26] Available in the website of UNFCCC. www.unfccc.int.

another 546 projects were pending registration with over 4 million CERs expected by the end of 2012.

From the point of view of the total of 206 methodologies approved by sector, almost 80% (188) were concentrated in 6 sectors: renewable energies, manufacturing industries, transport, waste management, chemical industries, and energy demand.

Of the total expected CERs corresponding to registered projects, 65.6% corresponded to China, 10.2% to India, 3.6% to Brazil, 2.7% to South Korea, and 1.9% to Mexico. These five countries together account for almost 85% of the annual CERs generated. If we take into account the CERs actually issued up to that moment, we find that 60.9% corresponded to China, 14.1% to India, 8.8% to South Korea, 6.8% to Brazil, and only 1.6% to Mexico. However, this figure is strongly influenced by the size of the projects, given that if the number of registered projects is taken, 52.9% of them corresponded to China, followed by India with 18.3%, Brazil (4.2%), Vietnam (3.3%), and Mexico (2.8%). These five countries accounted for 81.5% of the projects. In this case, Korea, with only 1.5% of the projects, does not appear within the relevant percentages. The explanation lies in the relative size of the projects corresponding to Korea, mainly with regard to those that refer to HFCs.

Almost three quarters of the registered projects (73.2%) corresponded to renewable energies, followed by waste treatment (11.7%). However, this does not necessarily correspond to the importance of these projects in the total number of CERs issued. Most of the CERs, in fact, correspond to projects related to HFCs, CH_4, and N_2O. Thus, priority is given to the attractiveness for business of the global warming potential (GWP)[27] of the gases, from the point of view of the magnitude of emissions that can be avoided in terms of carbon dioxide equivalent (CO_2e), more than other factors. It would be necessary to analyze whether the "discouragement" we were talking about earlier has anything to do with this.

As a detailed analysis of the information provided by the UNFCCC on its website and by the UNEP RISØ Centre (URC)[28] emerges, throughout this period since the emergence of the CDM, much more use has been made of the high GWP of certain gases than of the potential contribution to sustainable development of the projects or the Technology Transfer implicit in them. In this sense, the very "environmental integrity" of the mechanism as a whole would be called into question insofar as many of the projects do not generate real reductions in GHG emissions, as was the case with some projects related to the incineration of HFC-23 (obtained as a by-product of the production of HCFC-22), which generated more than half of the CERs issued

[27] The global warming potential (GWP) is a measure used to compare the effects of various GHGs on climate with respect to carbon dioxide (CO_2), taking into account the ability of a given quantity of each to increase in temperature in a given period compared to the same mass of CO_2. The unit of measurement used is the carbon dioxide equivalent (CO_2e).

[28] Ver el sitio Web de la CMNUCC www.unfccc.int y Fenhann (2008).

as of 2009.[29] Nor does the flow of Foreign Direct Investment (FDI) appear to be important, as many of the projects are financed by the local financial market.

More in-depth studies would have to be carried out to see whether there would be more incentives if the host countries of the projects were allowed to save these CERs so that they could be made available when it was most convenient for them (when their price in the various markets where they could eventually be traded was higher or when they might have to use them to meet any emission reduction commitments they might have to make at some point in time). Although this is not yet a high priority on the discussion agenda, at some point this issue will have to be defined.

There is also another point that is little talked about: the Kyoto Mechanisms have a raison d'être as long as the NAI Parties do not take on quantified emission reduction commitments. Otherwise, they would be faced with the paradoxical situation of delivering (or having delivered) at low cost their most accessible, cheap, and/or immediate mitigation options, leaving for them the most expensive and difficult to implement at the time when they would eventually have to assume a quantified commitment.

Another "perverse" point of the mechanism is that a situation is produced in which, the one who advances measures of mitigation, loses competitiveness from the point of view of the mechanism, because these reductions and/or limitations of emissions become part of its baseline. Thus, there is an incentive to delay the implementation of mitigation measures so that measures similar to these are additional and can be applied to the CDM. Additionally, an artificial differentiation is established between countries that have already carried out some of the less costly mitigation measures (fuel substitution, energy efficiency measures, introduction of renewable energies, etc.) and those that have not yet done so, generating an advantage for the latter.

Beyond the ethical issues inherent in the allocation of property rights over the environment (which is what the allocation of carbon emission permits is basically about), there is also the question of whether giving the market a key role in solving the problem of global climate change is not calling on the arsonist to help put out the fire. In reality, we have reached the current situation not because of a lack of market, but because of an excess of it. From the economic point of view, this is an externality

[29] https://ecologistasenaccion.org y http://www.multilateralfund.org/72/Spanish/1/S7241.pdf. HCF-23 (tri-fluoro-methane) is an unavoidable product of HCFC-22 (chloro-di-fluoro-methane). HCFC-22 was, until recently, one of the most widely used refrigerant gases, not only in domestic but also in industrial installations and as a foaming agent for extruded polystyrene, but it has gradually been replaced by its characteristic of being an ozone depleting substance (ODS), with a positive ozone depleting power (ODP). It is still manufactured in only a few countries: China, India, Argentina, North Korea, South Korea, Mexico, and Venezuela. At one time HFC-23 was recovered and used as a raw material for the production of certain products used as fire extinguishers (mainly Halon-1301), but this practice has fallen into disuse. Today a small amount of HFC-23 is used as an extinguishing agent in certain processes (semiconductor manufacturing and cryogenic refrigeration), but the vast majority of the HFC-23 produced is not consumed and is released into the atmosphere, captured, or destroyed (by incineration). Both HCFC-22 and HFC-23 are GHGs, the latter with a 100-year GWP equivalent of 11,700. In some cases, it was suspected that the economic return generated from the sale of the ERCs generated by the incineration of the by-product (HFC-23) justified the installation of the main product (HCFC-22) plant.

accumulated over time by the excessive use of a common property by a few players without having compensated the rest of the owners of that resource for that abusive use. Economic theory proposes solutions that are not only related to the assignment of property rights, but fundamentally to the regulation of this activity. The problem is one of lack of regulations rather than lack of market freedom, which in fact led us to this situation.

That the market has serious limitations in solving this problem is clearly shown by the evolution of the values of the units allocated in the EU's Emission Trading Scheme (ETS), as shown by Estrada Oyuela (2008).[30] This situation was mainly related to the allocation of permits, for each country, far above the actual emissions that were being recorded, which led to an excess supply of permits and consequently to an abrupt fall in their value. The argument that the SO_2 emissions permit market in the USA and the ETS-EU work cannot be extrapolated to the international level, where countries do not recognize a higher authority to which they have delegated the power to police and apply sanctions as they did in these two systems (represented in one case by the EU itself and in the other by the Environmental Protection Agency—EPA). It is clear that in such a system if someone does not comply with the rules of the game and is not penalized, there is no incentive for the other actors to comply. If a small country does not comply, it will probably be sanctioned. Would the same thing happen if some of the biggest countries did not comply?

An important point to bear in mind is that there is not just one carbon market, but several. This is not trivial. In addition, CERs issued through the application of the CDM (Article 12 of the KP) will have to compete with Emission Reduction Units (ERUs) arising from JI (Article 6 of the KP), emission reductions that are channeled through voluntary markets, those channeled through the Emissions Trading mechanism (Article 17 of the KP), and all other instruments and modalities through which transactions can be carried out. The role that the originally called "Hot Air"[31] can play is no less important, both by lowering the prices of the other certificates offered and by reducing the need to go out and look in other markets for the emissions reductions that ICs need to comply with their commitments.

Permit values will also vary depending on various situations. CERs are likely to be worth less than other certificates because they are subject to greater uncertainties, greater needs for controls, and also have a longer maturation period. Even among the CERs themselves, values will vary if they correspond to emission limitation or carbon sequestration projects (in which case they have a shorter expiry time and are likely to be worth less in the market). The higher the degree of project progress

[30] We will come back to this issue later.

[31] The concept of the "Hot Air" arises from the international negotiations prior to the establishment of the KP. It basically refers to the situation of countries such as Ukraine, the Russian Federation, and other Eastern European countries whose Assigned Amounts (AAs) in Annex B of the Kyoto Protocol for the base year (1990 in most cases) were significantly higher than the emissions levels observed at the time of the signing of the KP (1997), so that these amounts were thought to be "inflated" with respect to reality in order, among other things, to favor the exchange of additional emissions permits that were being obtained at no cost.

within the CDM cycle, the more value these avoided/reduced/sequestered emissions will potentially have.

Perhaps the main problem is that, from the beginning, too many expectations were created, and it was thought that these mechanisms (mainly the CDM) would act as a kind of Robin Hood that would redistribute resources from the rich to the poor. So far it was a kind of Hood Robin that made a number of "brokers" and intermediaries rich, but had a very poor contribution to the sustainable development of host countries and the Transfer of Technology from ICs to DCs and LDCs.

Obviously, this article cannot pretend to exhaust this discussion which, on the other hand (at least in some aspects), seems to have just begun. If the Market Mechanisms are to be given a role in preventing climate change, it would be important to try to integrate CDM project activities with the needs for adaptation and/or the reduction of vulnerabilities to the expected impacts of climate change on host countries. It cannot be that the only relationship between CDM and adaptation is the contribution of 2% of the value of the CERs to the integration of a fund, feeding the paradox that the poor finance themselves to cover their emergencies. There was some hope in the Programmes of Activities (PoAs), but although they extended the range of possibilities a little, they did not solve the basic problems and presented other challenges that did not make them such an interesting instrument as they had originally been. It would be very interesting if some future development of the CDM could be related to a deeper process of collaboration and facilitation of the sustainable development and Technology Transfer processes.

In fact, it is undeniable that there is an income (rent) appropriated by Annex I countries through the use of the CDM, while there is a significant cost differential between what it would cost them to reduce the ton of CO_2e internally (through the implementation of domestic measures in their own territory) and what it actually costs them to access the CERs.[32] So far, giving content to the CDM so that it effectively contributes to the achievement of the UNFCCC objectives is still a pending task.

5.7 CDM: What Could Be Expected After Doha?

The first commitment period of the KP covered the period from 2008 to 2012. For this reason, it was important that at the end of that period, the mentioned meeting set the conditions for the operation of the CDM thereafter.

The 18th Conference of the Parties (COP18) to the United Nations Framework Convention on Climate Change (UNFCCC) and the 8th Conference of the Parties to the UNFCCC acting as the Meeting of the Parties (CMP8) took place in the city of Doha, Qatar, between November 26 and December 8, 2012. A number of documents were adopted at these meetings, the most important of which are (a) the document approving the amendment of the Kyoto Protocol for a second commitment period (FCCC/KP/CMP/2012/L.9), which will run from January 1, 2013, to December 31,

[32] Gobierno de la República Argentina (1999).

2020; (b) the document declaring the work of the Long-Term Cooperative Action Group (AWG-LCA) created at COP13 in Bali completed, closing the Bali Roadmap process; and (c) relating to the revision of the CDM rules.

In the latter case, it was decided to revise the CDM modalities and procedures in order to adopt the changes at CMP9, for which parties were invited to make contributions until March 25, 2013. These inputs and the recommendations of the CDM-EB were considered by the Subsidiary Body for Implementation (SBI) at its meeting in June 2013. Issues were raised regarding CDM governance, methodologies, and additionality. On the latter, the CDM-EB called for extending simplified modalities for the demonstration of additionality to small-scale projects and committed to work on simplifying and streamlining methodologies in the quest to reduce transaction costs. The 45th Session of the Subsidiary Body for Scientific and Technological Advice (SBSTA)[33] was identified to discuss the eligibility of carbon capture and storage (CCS) projects under the CDM with transport or storage in more than one country, as well as the creation of a global pool of CERs. However, these issues have not yet been admitted to the CDM. As far as the negotiations on Reducing Emissions from Deforestation and Forest Degradation (REDD) are concerned, they got bogged down, mainly because of issues related to the verification of emission reductions.[34]

5.8 The Paris Agreement, the Poor Results of COP25, Trump and Beyond …

Carbon markets have shown some difficulties in functioning in recent years, mainly due to issues linked to the economic recession and the fall in demand for the various certificates issued by the different existing mechanisms.[35] The European Union's Emissions Trading Scheme (ETS) is no exception to this situation. The prices of the certificates (EUAs)[36] remained in a range of EUR 4 to EUR 7 (USD 5 to USD 9), when they had reached EUR 13 (USD 18) a few years earlier.

In this context, from 2013 onward the CDM situation also deteriorated even further from what had previously been discussed.[37] There are several reasons for this, but the main one is the enormous fall in CER prices, which reached historic lows in 2013 and 2014. The value of the CERs in those years reached U$S0.51 (€0.37), when they had reached peaks of U$S20 and, for a few years, maintained values of U$S10. This fall in prices does not generate monetary incentives to attract investment and to achieve additional emission reductions through the use of this mechanism and has

[33] August 2016.

[34] *Monitoring, Reporting and Verification* process (MRV).

[35] GBD Network (2014). To this drop in demand must be added the exit from the second KP commitment period (2013–2020) of some countries that had made quantified commitments in the period 2008–2012: Canada, Japan, Russian Federation, New Zealand.

[36] Emission Units Allowances.

[37] See point 6.

led some countries (Mexico, for example) to not seek to present new projects and to only generate CERs from projects that are already registered.[38] The cause of this collapse in prices is to be found firstly in an oversupply of CERs compared to the low demand for them at the end of the first KP commitment period by countries that had to meet their quantified commitments. Another factor contributing to this fall is that the price of CERs is no longer linked to the value of the EUAs, as had been the case in the past, mainly due to the lack of sustained demand for CERs by EU-ETS participants.

In view of this situation both the European Commission and the UNFCCC are seeking to reformulate their respective instruments (EU-ETS and CDM). In the case of the EU-ETS, the European Commission proposed a short-term mechanism called "backloading" aimed at removing 900 million EUAs from the market to be reinjected after 2020,[39] with the aim of achieving a recovery in the price of certificates.[40] As far as the CDM is concerned, the greatest expectations of saving the mechanism lie in the possibility of including carbon capture and storage (CCS) projects and the new Emissions Trading Schemes that would potentially be established in countries such as South Korea, China, and at subnational levels in the USA and Canada, as well as in the Carbon Offset and Reduction Plan in the International Aviation sector, approved by the International Civil Aviation Organization (ICAO) to come into operation in 2021. But mainly on the possibility of using CERs to comply with NDCs in order to increase the ambition of the commitments made in the context of the post-2020 PA.

According to the latest annual report of the CDM Executive Board (CDM-EB),[41] the mechanism has already registered more than 8000 projects in 111 countries, Annex I Parties have used more than 1000 million CERs to meet their KP commitments, and about 1900 million CERs have been issued to be used for the fulfillment of the KP second commitment period (or for any other use, including the fulfillment of commitments in the PA if it is established that this may be possible). If so, the current situation of very low demand for CERs and a high level of uncertainty in the mechanism that led, for example, to some projects ceasing to issue the corresponding CERs could be reversed. In this respect, 41% of the projects for which CERs had been issued up to the end of the first commitment period of the KP (31/12/2012) ceased to do so after that date.[42] This situation also led to an impact on the entities in charge of validating and verifying projects developed under the CDM.[43]

In the context of the PA, many parties have formulated their NDCs on the basis of access to and the possibility of participating in an international carbon market. On this point, there is an ongoing discussion on how to implement the "cooperative approaches" to the implementation of NDCs set out in Article 6 of the PA. While some argue for a centralized global market with UNFCCC oversight (similar to that

[38] Carbosur (2014) and Gobierno de México (2018).

[39] Carbosur (2014) and GBD Network (2014).

[40] They estimate to arrive at a price of €7.5 for the EUAs. See GBD Network (2014).

[41] UNFCCC (2017).

[42] Ibid.

[43] Ibid.

under the KP), others argue for a less rigorous approach from the point of view of centralized oversight. In this discussion, the role that the CDM can play in a post-2020 context is unavoidable. One topic of discussion is whether the CERs will be able to be used to meet the commitments arising from the PA. The large supply of certificates at very low prices raises concerns about whether their continued use could undermine new mitigation initiatives. However, if there is no CDM, there is a risk that the capacity, experience, and knowledge required will be lost. Once lost, these attributes will not be available to support the international carbon market in the future and will have to be re-established, perhaps at a much higher cost than it took to develop them.

An additional factor of uncertainty was the announcement by the President of the USA of his intentions to "withdraw" his country from the PA,[44] while a good dose of expectation in the recovery of the CDM is placed in the role of subnational mechanisms of Emissions Trading, among which there is a special participation that may be developed among some of its states, in some cases, interacting with some Canadian Provinces. So far there is no definition of this, although some facts play in favor of the "integrity" of the international climate regime. In a sense, none of the Parties is willing to "take their feet off the plate", aware of the difficulty of negotiating another agreement. Added to this is the inertia inherent in agreements, in the sense that both countries and companies need to make strategic investment decisions that require them to do so with a certain amount of anticipation and, once they have embarked on such plans, it is not easy to go back. Nor was there a "domino effect" in the sense of statements or attitudes from the rest of the "Parties" to the climate agreements for following in President Trump's footsteps. And, to complete, China's role so far has been very "proactive" in the sense of ratifying its commitment to comply with what was agreed.[45]

Seeking to achieve the necessary agreements to meet the goal of avoiding temperature increases over 1.5 °C, COP25[46] was extended two days longer than planned. However, it ended without achieving significant results either with regard to the objective of setting more ambitious commitments to limit and/or reduce GHG emissions or regarding to the goal of establishing a consensus to regulate carbon markets.[47] In spite of those poor results, an agreement was reached called "Chile-Madrid. Time for Action", which asks for increasing the "climate ambition" in 2020 and complies with the Paris Agreement seeking that countries submit more ambitious NDCs to respond to the "climate emergency", seeking to close the "emissions gap" between

---

[44] Announcement made on June 1, 2017. See BBC News https://www.bbc.com/mundo/noticias-internacional-40124921.

[45] See http://www.telam.com.ar/notas/201706/191078-cambio-climatico-acuerdo-paris-trump-eeuu-china-europa.html.

[46] Organized by Chile, but held in Madrid (between December 2 and 15, 2019) as a result of the social conflicts and protests that began in mid-October 2019 in Chile. It was originally scheduled to take place between December 2 and 13, 2019, but was extended by two days in the search for stronger agreements.

[47] BBC World News (2019).

the current state of GHG emissions and what science recommends to avoid significant effects as a result of climate change. Nevertheless, a number of countries did not adhere to this commitment, including the USA, China, India, Saudi Arabia, Japan, Brazil, and Australia. COP26 will be held in Glasgow at the end of 2020.[48]

Therefore, no significant changes in the conditions and characteristics of the CDM and in carbon markets architecture are foreseen, which would modify the main aspects indicated above. More than 20 years after the KP, the search for a consensual, binding, and ambitious international climate agreement continues. However, substantive decisions continue to be postponed. In short, it is impossible for a market to survive without demand. And the demand arises from the degree of rigor that exists in the fulfillment of the commitments that are assumed and to what extent these commitments imply a real effort to reduce emissions.

References

BBC World News (2019) COP25: three keys of the controversial new climate agreement, 15 Dec 2019. https://www.bbc.com/mundo/noticias-internacional-50800493

Bouille D, Girardin LO et al (1999) Study on flexibility mechanisms within the context of the UNFCCC and the KP. MRECIC-Argentina, Ministry of Environment of Canada, World Bank, Buenos Aires

Bouille D, Girardin LO, Di Sbroiavacca N (2000) Argentina case study. In: Biagini B (ed) Confronting climate change. Economic priorities and climate protection in developing nations. NET, Pelangi, Washington

CAEMA (2003) The state of development of national CDM offices in Central and South America. CAEMA, Department of Foreign Affairs and International Trade, Climate Change and Energy Division, Canada

Carbosur (2014) The situation of the carbon market. Montevideo. http://www.carbosur.com.uy/article/la-situacion-del-mercado-de-carbono/

Criqui P, Kouvaritakis N (1997) The costs for the energy sector of reducing CO_2 emissions: an international assessment with the POLES model. In: Research notebook No. 13. IEPE, Grenoble University of Social Sciences

Estrada Oyuela R (2008) The carbon certificates market. In: CEI magazine—foreign trade and integration, No. 11, Buenos Aires, May 2008

Fenhann J (2008) CDM pipeline, June 2008. UNEP Risø Centre, Dinamarca

Figueres C (ed) (2002) Establishing national authorities for the CDM. A guide for developing countries. CSDA, CCKN, IISD, Winnipeg

GBD Network (2014) Carbon market. Current situation and expectations of COP20. In: Outlook 2020. Foresight. Member of GBD Network. Report No. LXXVIII, July 2014

Girardin LO (2000) Global climate change and the distribution of mitigation costs of its eventual consequences among different countries. Buenos Aires. Available at http://fundacionbariloche.org.ar/

Girardin LO (2008a) Opportunities and challenges for taking advantage of the CDM. Presentation to the seminar "clean development mechanism of the Kyoto Protocol". Montevideo, Uruguay, 9 Sept 2008

Girardin LO (2008b) Regional impacts associated with climate change. Case study: southern cone of South America. Bariloche Foundation, Sustainable Southern Cone, Buenos Aires

[48] UN News (2019).

Girardin LO (2013) Socioeconomic and political aspects of climate change. From the convention to the Kyoto Protocol (1990–2000). Patagonia Third Millennium Foundation, Trelew-Buenos Aires. ISBN 978-987-26155-8-1. Available at http://www.patagonia3mil.com.ar/wp-content/uploads/libros/publicaciones-politicas_ambientales.pdf
Girardin LO (2018) Myths and realities of the role of the CDM and other market mechanisms in their contribution to sustainable development. Rev Sci Res 68(5):72–86. Available at https://aargentinapciencias.org/publicaciones/revista-resenas/revista-cei-tomo-68-no-5-2018/
Girardin LO, Bouille D (2002) Learning from the Argentine voluntary commitment. In: Baumert K et al (ed) Building on the Kyoto Protocol: options for protecting the climate. WRI, Washington
Girardin LO, Bouille D (2003) Conditions for greater commitment of developing countries in the mitigation of climate change. CCKN, IISD, Winnipeg
Girardin LO, Di Sbroiavacca N (2000) México case study. In: Biagini B (ed) Confronting climate change. Economic priorities and climate protection in developing nations. NET, Pelangi, Washington
Gobierno de México (2018) Mecanismo de Desarrollo Limpio (MDL). Instituto Nacional de Ecología y Cambio Climático (INECC), Acciones y Programas. Available at https://www.gob.me/inecc/acciones-y-programas/mecanismo-de-desarrollo-limpio-mdl
Government of the Argentina Republic (1999) Study on flexibility mechanisms within the context of the United Nations Framework Convention on Climate Change and the Kyoto Protocol. Final report, Nov 1999. Ministry of Foreign Affairs, International Trade and Worship, Environment Canada, World Bank, Buenos Aires
Herzer H (1990) Disasters are not as natural as they seem. Environ Urban (30):3–10. IIED-AL, Buenos Aires
http://www.unfccc.int
Intergovernmental Panel on Climate Change-IPCC (1998) The regional impacts on climate change. In: A special report of IPCC Working Group II. Cambridge University Press, London
Intergovernmental Panel on Climate Change-IPCC (2007) Climate change 2007: climate change impacts. Adaptation and vulnerability. In: Working Group II contribution to the IPCC 4AR. Cambridge University Press, London
Intergovernmental Panel on Climate Change-IPCC (2014) Climate change 2014. Impacts, adaptation and vulnerability. In: Working Group II contribution to the IPCC 5AR. Cambridge University Press, London
Lipietz A (1995) Enclosing the global commons: global environmental negotiations in a north-south conflictual approach. In: Bhaskar V, Glyn A (eds) The north, the south and the environment. Ecological constraints and the global economy. UNU, Tokyo
Natenzon C, Murgida AM, Ruiz M (2006) Social vulnerability to probable climate change. In: Serman & Asociados (ed) Socioeconomic impacts of climate change. Document prepared for the second national communication from the government of Argentina to the parties to the UNFCCC. Buenos Aires
UN News (2019) COP25 ends with little progress in terms of reducing carbon emissions, 15 Dec 2019. https://news.un.org/es/story/2019/12/1466671
United Nations Framework Convention on Climate Change (1992) United Nations Framework Convention on Climate Change. Available in Spanish at https://unfccc.int/files/essential_background/background_publications_htmlpdf/application/pdf/convsp.pdf
United Nations Framework Convention on Climate Change (1997) Kyoto Protocol. Available in Spanish at https://unfccc.int/resource/docs/convkp/kpspan.pdf
United Nations Framework Convention on Climate Change (2006) Impacts vulnerabilities and adaptation to climate change in Latin America. Background paper. UNFCCC Secretariat, Bonn
United Nations Framework Convention on Climate Change (2007) Climate change: impacts, vulnerabilities and adaptation in developing countries. Bonn
United Nations Framework Convention on Climate Change (2015) Paris Agreement. Available in Spanish at https://unfccc.int/files/meetings/paris_nov_2015/application/pdf/paris_agreement_spanish_.pdf

United Nations Framework Convention on Climate Change (2017) Annual report of the executive board of the clean development mechanism to the conference of the parties serving as the meeting of the parties to the Kyoto Protocol. FCCC/KP/CMP/2017/5. Bonn

World Resources Institute (2016) When can the Paris Agreement come into force? Interactive map. http://www.wri.org/blog/2016/04/when-could-paris-agreement-take-effect-interactive-map-sheds-light-es

Chapter 6
Regional Study on the Economics of Climate Change in South America. Argentine Chapter (ERECCS-Argentina). ECLAC

Abstract This chapter provides a brief summary of the results of the Regional Study on Economics of Climate Change in South America (ERECCS) for Argentina, presenting the main results obtained in the attempt of monetary valuation of the expected impacts of climate change and the measures, both for Adaptation and Mitigation that were identified throughout the development of the study.

Keywords South America · Climate change economics · Mitigation costs · Climate change impacts costs · Adaptation costs

The Study on the Economics of Climate Change in Argentina was part of a regional initiative whose main objective is to demonstrate the economic importance that climate change can have for societies, productive systems, and the natural heritage of the countries in the Region, in this particular case, the Argentine Republic. In this way, decision makers at the national and local levels will be able to have a tool that allows them to take into account, in their analysis, the relevant costs, and benefits.

This work carried out for Argentina, in which a series of experts and researchers from various universities and research centers (both public and private) and also from government bodies participated, constitutes a first approximation to the monetary valuation of the expected effects of climate change and of the different aspects of vulnerability to this phenomenon presented by the sectors, systems, and regions of the country that were analyzed in this study, as well as the adaptation and mitigation measures identified, offering information and concrete data for decision-making. Some of these aspects were analyzed in the National Communications on Climate Change and in this work they were deepened. Others were identified at that time as objectives for future studies and were addressed in this case. However, much work remains to be done in terms of extending the coverage of this analysis to other economic sectors, geographic regions and systems, both natural and human, which were not analyzed in detail on this occasion.

Leonidas Osvaldo Girardin (Coordinator of the Study).

L. O. Girardin, *Socioeconomic and Geopolitical Aspects of Global Climate Change*, The Latin American Studies Book Series, https://doi.org/10.1007/978-3-031-53246-7_6

The main precedent in studies of this type is the Stern Report, and an attempt was made to replicate it by adapting it to the reality of the Region; in the sense that, beyond the effort made in terms of mitigation and the effectiveness it may have, the main cost of not acting is given by the weight of the expected impacts and the vulnerability to climate change and variability.

The impacts and corresponding adaptation measures were obtained from Climate Scenarios developed by INPE Brazil, based on the IPCC's A2 and B2 Emission Scenarios.

In order to develop the study, it is necessary to establish a socioeconomic and demographic scenario that serves as a basis for subsequent projections of the variables that explain the evolution of the country's economic activity in the 2005–2100 period, while defining the macroeconomic scenario used throughout the study. This scenario also takes into account the effects on the country of both the international context and the relationship between economic development and emissions in the sectors analyzed and presents a greater degree of disaggregation between 2005 and 2030 and projections until 2100. The definition of these scenarios is also important for the identification of those factors (drivers) that serve for the development of the predictive model of the energy situation and the analyses aimed at determining the future behavior of the other sectors that emit greenhouse gases (GHG).

The impacts taken into account in the study are as follows:

Box 1 Regions and sectors analyzed. Impacts and indicators identified

Flows in the Comahue Region (provinces of Río Negro and Neuquén)	Loss of income due to flow reduction in hydroelectric power plants
Flows in the Cuyo Region (Mendoza and San Juan Rivers)	Social cost of water in predicted situations of water stress based on the sectoral report of water supply and demand in the provinces of Cuyo
Flows in the Coastal Region (Paraná and Uruguay Rivers)	Potential economic loss due to the decrease in hydroelectric generation as a consequence of the anticipated reduction in long-term flows
Agriculture	Impact of the variation of the productivity of the main crops studied (wheat, corn, and soybean)
Forests of Northwest Argentina (NOA)	Loss of environmental goods and services as a result of deforestation (i) water resources protection services; (ii) climate regulation services; (iii) flood and extraordinary flood prevention services; (iv) soil formation services; (v) biological control services; (vi) recreation related services; (vii) provision of natural medicines; (viii) replacement costs; and (ix) biodiversity losses in terms of flora

(continued)

(continued)

Box 1 Regions and sectors analyzed. Impacts and indicators identified

Iberá Marshlands	Loss of environmental goods and services as a result of deforestation (i) water resources protection services; (ii) climate regulation services; (iii) flood and extraordinary flood prevention services; (iv) soil formation services; (v) biological control services; (vi) recreation related services; (vii) provision of natural medicines; (viii) replacement costs; and (ix) biodiversity losses in terms of flora
Human health	Costs of potential increase in dengue and malaria cases
Río de la Plata coastline	Potential economic impact by flooding of: (i) public service infrastructure; and (ii) building infrastructure
Litoral Region (Uruguay and Paraná Rivers)	Monetary value of losses caused by prolonged flooding of the region's rivers, taking into account factors such as the number of cases and the duration of each episode

Source Adapted from CEPAL (2014) and Girardin (2017)

For their part, the adaptation measures identified have been the following:

Box 2 Valued adaptation measures

Region or phenomenon	Measures
Flooding in the Rio de la Plata	Two types of measures were considered in order to determine their cost: (i) construction of defenses and (ii) relocation of human settlements to non-flood areas
Flooding in the rivers of the Litoral (Paraná and Uruguay)	The following measures were taken into consideration: (i) construction of defenses and (ii) emergency and evacuation measures taken in previous cases
Comahue Region	The cost of using water for irrigation of the hectares planted with fruit trees was taken into consideration
Human health	Adaptation measures consisting of dengue vector control, surveillance activities, and services were considered

Source Adapted from CEPAL (2014) and Girardin (2017)

The results of the respective monetary valuations were converted to 2005 dollars, and various discount rates were applied. The calculation was applied to both the projected impacts of climate change and the adaptation measures identified. The costs of mitigation measures were also estimated by applying different discount rates.

To determine the magnitude of these costs, they were calculated as percentages of Argentina's GDP in the base year (2005) according to the results as shown in Tables 6.1, 6.2, and 6.3.

Table 6.1 Costs of expected climate change impacts

Summary of the forecasted weighted cost of climate change impacts					
(As a percentage of 2005 GDP)					
Scenario and discount rate	2020	2030	2050	2070	2100
A2 0%	7.13	10.99	10.05	9.21	144.35
B2 0%	6.27	10.66	12.02	6.91	127.41
A2 0.5%	6.83	10.29	9.39	8.47	98.85
B2 0.5%	6.00	9.97	11.00	6.90	87.35
A2 2%	6.00	8.53	7.77	6.99	34.48
B2 2%	5.28	8.22	8.63	6.51	30.85
A2 4%	5.07	6.75	6.20	5.75	11.59
B2 4%	4.48	6.46	6.54	5.65	10.78

Source Adapted from CEPAL (2014) and Girardin (2017)

Table 6.2 Identified adaptation costs. Scenario A2

Accumulated economic cost of measures to adapt to climate change identified					
A2 scenario					
(As a percentage of 2005 GDP)					
Discount rate	2020	2030	2050	2070	2100
0%	0.96	1.68	2.55	3.89	5.76
0.50%	0.93	1.58	2.32	3.34	4.59
2%	0.85	1.34	1.79	2.24	2.63
4%	0.75	1.10	1.34	1.50	1.59

Source Adapted from CEPAL (2014) and Girardin (2017)

Table 6.3 Identified adaptation costs. Scenario B2

Accumulated economic cost of measures to adapt to climate change identified					
B2 scenario					
(As a percentage of 2005 GDP)					
Discount rate	2020	2030	2050	2070	2100
0%	0.80	1.40	2.33	3.37	5.2
0.50%	0.78	1.32	2.10	2.90	4.12
2%	0.71	1.12	1.59	1.95	2.33
4%	0.64	0.93	1.17	1.30	1.38

Source Adapted from CEPAL (2014) and Girardin (2017)

Table 6.4 Identified mitigation costs

Total accumulated gross cost of identified mitigation measures					
Both scenarios					
(As a percentage of 2005 GDP)					
Discount rates	Years				
	2020	2030	2050	2070	2100
Totals (4%)	5.98	22.67	80.05	234.41	653.49
Totals (12%)	11.08	42.81	152.09	332.74	827.66

Source Adapted from CEPAL (2014) and Girardin (2017)

The analysis of the preceding tables shows the influence that the discount rate used has on the monetary valuation of the impacts and measures and, consequently, on how they will be ordered in terms of their value, independently of the impact in physical terms (very important but very distant effects in time can be valued relatively less than smaller but closer impacts, depending on the discount rate used).

The results of the estimated costs of the mitigation measures analyzed are as shown in Table 6.4.

As can be seen, the total cost of the mitigation measures calculated in this study considerably exceeds the estimated cost of the identified impacts and the cost of the adaptation measures analyzed and assessed. However, these results should be interpreted as a reflection of the greater ease of collection of information needed to calculate emissions savings and mitigation costs and the certainty with respect to this information, taking into account the general difficulties that exist, both in terms of the availability of information on impacts and possible adaptation measures, the greater uncertainty about these data, and the additional difficulty of calculating the economic cost of goods and services, especially those provided by the environment, to which a market price cannot easily be assigned.

(*) The study is available on the website of the United Nations Economic Commission for Latin America and the Caribbean (ECLAC) (in the format of a Synthesis Report) and on the website of Fundación Patagonia Tercer Milenio (the Synthesis Report plus all the Sectoral Reports):

http://www.cepal.org/cgi-bin/getProd.asp?xml=/dmaah/noticias/noticias/6/52296/P52296.xml&xsl=/dmaah/tpl/p1f.xsl&base=/dmaah/tpl/top-bottom.xsl

http://patagonia3mil.com.ar/publicaciones/

References

Comisión Económica de las Naciones Unidas para América Latina y El Caribe—CEPAL (2014) La Economía del Cambio Climático en la Argentina. Primera aproximación. Santiago de Chile, 241 pp

Girardin LO (2017) La Economía del Cambio Climático en la Argentina. Tomo I: Informe de Síntesis e Informe sobre Valorización Económica. Comisión Económica de las Naciones Unidas para América Latina y El Caribe (CEPAL), Consejo Nacional de Investigaciones Científicas y Tecnológicas (CONICET), Fundación Patagonia Tercer Milenio (FUNPAT), Buenos Aires, Trelew, 366 pp. ISBN 978-987-45525-3-2. Retrieved from http://www.patagonia3mil.com.ar/publicaciones/

Chapter 7
Forest Fires in Australia: Are We Inevitably "In the Oven" Also in Argentina?

Abstract This chapter describes Australian forest fires, occurred last south-hemisphere summer (2019/2020) that had catastrophic effects on vegetation, wildlife, environment, and economy and how these fires could be an useful experience to Argentina. Although forest fires in Australia are common and generally have natural causes—although it cannot be ruled out that some events could be intentional—this time these events have some unique characteristics: (a) they started much earlier than usual; (b) occur in a very particular local, regional, and global climate context that combines extreme heat, prolonged drought, and strong winds with factors linked to the dynamics of the climate in the Indian Ocean and a greater probability of occurrence and magnitude of forest fires globally; (c) economic issues (agribusiness); and (d) political factors, that prioritize specific sectoral interests. Major worse evils can always be avoided when society is prepared. The challenge for Argentina is to be prepared so that the forest fires that occur in different regions of the country at different times of the year do not generate catastrophic effects. For this, it is essential to have adequate strategies and policies. Climate change does not help, but even less inaction, unlearning, and neglecting to think that everything is magically solved by the market.

Keywords Forest fires · Extreme events · Policies and measures · Economic costs · Socioeconomic drivers

The year 2020 began with some terrible news from an environmental perspective: Australia is experiencing one of its worst bushfire seasons, with all that this fact means in terms of the loss of unique biodiversity in the world. Although its geographical distribution is quite extensive, it is mainly affecting the Eastern States, such as New South Wales and Victoria (where Sydney and Melbourne are respectively located), so the number of the total population affected is also very significant.

It is difficult for anyone, whether or not they are connected to environmental issues, to remain indifferent to a situation that is presented as a catastrophe from both the

---

Leónidas Osvaldo Girardin, Economist, CONICET/Fundación Bariloche.

L. O. Girardin, *Socioeconomic and Geopolitical Aspects of Global Climate Change*, The Latin American Studies Book Series, https://doi.org/10.1007/978-3-031-53246-7_7

natural and human points of view. In this sense, there are already dozens of deaths and a similar number of disappearances; more than 1500 homes destroyed[1] (other sources mention 15,000[2]), at least 140,000 people evacuated, including permanent residents and tourists (counting the State of Victoria alone),[3] and a budget of around $1.4 billion to alleviate the crisis.[4] Canberra has become one of the most polluted cities in the world,[5] as a result of the smoke from the fires (whose influence has already reached even Argentina, Chile, and Uruguay). In terms of the impacts on productive activities, there are significant losses in the agricultural and livestock sectors, numerous power cuts have been caused, and in some locations fuel and food shortages have begun to be experienced.[6]

Biodiversity and ecosystem losses are even more difficult to quantify. There is speculation that the number of dead animals has already reached one billion[7] (it was already between 480[8] and 500[9] million at the beginning of January), and particular emphasis is placed on the loss of populations of endangered species such as koalas (it is estimated that between 8000[10] and 25,000[11] have disappeared), which are closely linked to the dynamics of natural eucalyptus forests. However, not only is the problem of the disappearance of animals a direct consequence of the fires, but there are also the short-, medium-, and long-term effects of the destruction of their habitats, leaving them vulnerable long after the fires have been extinguished. In this sense, some sources[12] state that 1.25 billion animals are affected, including the dead, the injured and those whose habitats were significantly damaged.[13] Considering that 80% of Australia's animals are endemic to the island[14] (and in many cases unique), we have a better idea of the impact that these fires can have on the world's wildlife.

---

[1] See Telesur (13/01/2020) https://www.telesurtv.net/news/australia-incendios-forestales-millones-hectareas-quemadas-20200113-0020.html and also Infobae (07/01/2020) https://www.infobae.com/america/mundo/2020/01/07/por-que-los-incendios-en-australia-son-tan-devastadores-y-no-se-detienen-tres-claves-que-explican-la-dramatica-situacion/.

[2] See Página 12 (07/01/2020) https://www.pagina12.com.ar/240276-las-causas-del-incendio-por-que-australia-vive-su-propio-inf.

[3] See Página 12 (op. cit.).

[4] See Página 12 (op. cit.).

[5] See Milenio Digital 2020 (20/01/2020) https://www.milenio.com/internacional/europa/la-sequia-el-aliado-de-los-incendios-en-australia.

[6] See C/NET en Español (13/01/2020) https://www.cnet.com/es/noticias/australia-incendios-causas-consecuencias-koalas/.

[7] See Telesur (op. cit.).

[8] See Página 12 (op. cit.).

[9] See Sin permiso (04/01/2020) http://www.sinpermiso.info/textos/australia-esta-cometiendo-su-suicidio-climatico and also Infobae (op. cit.).

[10] Telesur (op. cit.).

[11] Página 12 (op. cit.).

[12] See Telesur (op. cit.).

[13] Pictures of firefighters and volunteers throwing carrots and other tubers from helicopters to feed surviving wild animals had been seen. See Telesur (op. cit.).

[14] See Telesur (op. cit.).

To give an idea of the magnitude of what is happening, the more than 200 fire spots that were active in mid-January have consumed, according to some sources, more than 10 million ha,[15] and even the most conservative estimates maintain that at least 6 million ha[16] are affected. This means, at least, between more than double to four times the area affected by the fires that occurred, during the year 2019, in the Amazon (2.5 million ha).

Wildfires are common in Australia and generally involve natural causes (such as lightning strikes), although it cannot be ruled out that some outbreaks, this time, may have been intentional.[17] However, these fires in the season 2019–2020 have some particular characteristics. Firstly, they started much earlier than usual; in late September, early October 2019 (when, generally speaking, the bushfire season in Australia starts in January) and could probably last until March (which is when, generally speaking, these processes take place), when temperatures drop and regular rainfall starts.[18] Furthermore, these fires occur in a very special climatic context, in which various factors combine to make the situation worse: (a) Extreme heat (2019 was the hottest year on record for Australia, with a day on 18 December with an average temperature of 41.9 °C, a historical record); (b) Prolonged drought (with very little rain since 2017, making it the driest season in the last 120 years, and with a forecast of 3 drier and hotter than average summer months); and (c) Strong winds (up to 96 km/h).[19]

These local climatic conditions (extreme heat, prolonged drought, strong winds), converge with others that are regional (linked to the dynamics of climate in the Indian Ocean) and, of course, with those of a global nature (related to the influence of climate change on the causes that affect the greater probability of occurrence and greater magnitude of forest fires). In any case, there are also socio-economic and political determining factors.

Empirical evidence shows that it is getting hotter and hotter in Australia.[20] The average annual temperature from 1970 to date is about 1 °C above the average 1961–1990, while between 1910 and 1970 it was about 1 °C below that average. The IPCC Assessment Reports in 2007 (4AR) and 2013 (5AR) already warned of both heat waves and prolonged droughts and of the increased likelihood of favorable conditions (hot winters with their impact on soil moisture loss, low rainfall) for more frequent and intense wildfires.[21] Drier soils, more heat, less rain, result in more stressed forests and more fuel for fires.[22]

---

[15] Telesur (op. cit.).

[16] See Infobae (op. cit.).

[17] See C/NET (op. cit.).

[18] See Infobae (op. cit.), C/NET (op. cit.) and also BBC Mundo News (06/01/2020) https://www.bbc.com/mundo/noticias-internacional-50992270.

[19] See C/NET (op. cit.), BBC Mundo News (06/01/2020, op. cit.) and also Infobae (op. cit.).

[20] See BBC Mundo News (06/01/2020, op. cit.), IPCC (2007, 2014).

[21] See BBC Mundo News (06/01/2020, op. cit.).

[22] See C/NET (op. cit.).

However, in addition to these local environmental conditions, there are others that influence a wider geographical area, such as the Indian Ocean Dipole, commonly called the "Indian El Niño".[23]

This regional phenomenon is explained by the difference in sea surface temperatures in opposite parts of the Indian Ocean. The "Indian El Niño" has three types of phases: positive, negative, and neutral. In each of these phases, the temperatures in the eastern part (Oceania) range from warm to cold with respect to the western part (Africa). This Dipole 2019/2020 is in a positive phase, which means that in the western region of the Indian Ocean the temperature is much higher than normal and that in the eastern region it is much colder than usual. This positive phase, in turn, is much stronger than those experienced in the previous 6 decades (noting that extreme events occurred in 1961, 1994, and 1997)[24] and caused Floods and Landslides in East Africa (with precipitation 300% higher than the average for that area), as well as Drought, Heat Waves, and Wildfires in Australia and Southeast Asia. In these episodes, rain tends to move with the warm waters (where it rains more than average), while in areas with colder waters it rains less than average and the temperature is higher than average (this summer in Australia there were several days with temperatures above 40 °C), as a result of the lack of rain.[25]

As far as global conditions are concerned, although the responsibility for climate change is very clearly identifiable in the impacts on coral reefs or giant kelp forests (from the increase in water temperature at the surface of the oceans and the acidification of the oceans), concentration, climate change could also worsen the effects of the Indian Ocean Dipole, because extreme weather events could become more common in the face of rising concentrations of Greenhouse Gases (from one every 17.3 years to one every 6.3 years).[26] Furthermore, from the point of view of the local environment, climate change creates the conditions for more intense and more frequent forest fires, which in turn not only increase CO_2 emissions (it is estimated that in 3 months 350 million tons of CO_2 were released from forest fires; which, according to some estimates, may take up to a century to be absorbed naturally) but also the "smoke clouds" (pyrocumulus) that form can generate their own microclimate and trigger, for example, thunderstorms or whirlwinds.[27]

With regard to these characteristics of forest fires, there is talk of "sixth generation" fires, a type that is so powerful and intense that by itself it modifies climatic conditions, that can change direction unexpectedly, that have such a voracity that they

---

[23] The IOD is defined as a "Large-scale mode of inter-annual variability of sea surface temperature in the Indian Ocean. This pattern manifests through a zonal gradient of tropical sea surface temperature, which in one extreme phase in boreal autumn shows cooling off Sumatra and warming off Somalia in the west, combined with anomalous easterlies along the equator." IPCC (2013) IPCC Fifth Assessment Report (5AR). Glossary. https://www.ipcc.ch/site/assets/uploads/2018/08/WGI_AR5_glossary_ES.pdf.

[24] See BBC Mundo News (09/12/2019) https://www.bbc.com/mundo/noticias-50705069.

[25] See BBC Mundo News (09/12/2019, op. cit.).

[26] See BBC Mundo News (09/12/2019, op. cit.).

[27] See BBC Mundo News (06/01/2020, op. cit.).

can wipe out everything in their path[28] and that, according to certain sources, first appeared in Portugal and Chile in 2017.[29,30] The reasons for these fires are not always the same: In the USA (California, 2018) and in these 2019–2020 in Australia, the origin has much to do with the consequences of the combination of certain climatic and environmental conditions (prolonged droughts, intense, and more frequent heat waves), while in the cases of Brazil (Amazon, 2019) or Indonesia (2019), the causes are much more related to economic issues linked to both agricultural and forestry businesses and other economic interests.[31] However, in some cases, these events have points in common, when environmental issues, mainly climatic (prolonged droughts, heat waves), are combined with aspects linked to patterns of economic exploitation of natural heritage that prepare the ground for forest fires to find optimal conditions for expansion (intensive land use, monoculture, replacement of the original forest, search for the highest possible profitability in the shortest time).[32] In the case of the fires in Galicia (2017) and the already mentioned fire in Portugal (2017), pine and eucalyptus plantations were considered as part of the problem of the proliferation of fires, as many experts judge as the most effective to spread forest fires, and even have the particularity of growing back well after the fire.[33]

The fact that this catastrophe occurs in an industrialized country with a high standard of living and significant environmental awareness on the part of a large part of its population is a wake-up call. It is also proof of a structural crisis of consumption and production patterns, predominant until now, which are not sustainable in the long term in this context of a crisis, both environmental and socioeconomic, which is not necessarily assumed by those who have the obligation to make the decisions to face it in the best way. Such environmental crisis is one in which the climate crisis is an important component (but not the only one); and a socioeconomic crisis in which the consequences of inequality are increasingly evident (both in the appropriation of the benefits of resource exploitation, as well as in the suffering of the liabilities that this exploitation generates). In this sense, the internal criticisms received by the Australian Prime Minister are related to his statements regarding the lack of responsibility for climate change in the fires, his defense of the interests of the coal industry and his physical absence at the worst moment of the fires.[34] It is not idle to recall that Australia has high per capita emissions and that, within the group of developed countries, it is not one of the most committed to deepening international

---

[28] See El País (op. cit.).

[29] See El País (op. cit.).

[30] Others classified Chile's fires as belonging to "fifth generation". See Telam (op. cit.).

[31] See El País (op. cit.).

[32] See Página 12 (op. cit.).

[33] See El Independiente (06/07/2018) https://www.elindependiente.com/desarrollo-sostenible/2018/07/06/guerra-al-eucalipto-el-polvorin-de-los-incendios-en-galicia/, El Periódico (16/10/2017) https://www.elperiodico.com/es/medio-ambiente/20171016/incendios-galicia-portugal-coctel-calor-sequia-eucaliptos-6356310 and also Libertad Digital (20/06/2017) https://www.libertaddigital.com/internacional/europa/2017-06-20/el-incendio-perfecto-de-portugal-eucaliptos-desertificacion-y-bombas-de-fuego-1276601388/.

[34] See Página 12 (op. cit.) and Sin permiso (op. cit.).

negotiations on climate change, precisely because of its "national circumstances", among which is the significant importance of the coal sector in its economy and productive matrix.

The immediate question that arises is: If this happens in a "first world" country (often taken as a model of what "we could have been and were not" by our more conservative historiography), what awaits the inhabitants of countries like Argentina, which are in a process of laborious exit from the economic and social crisis in which we were submerged by the neo-free-trade theory[35] characterized by multiple restrictions and a state that increasingly disengaged from their responsibilities of control, monitoring, and punishment.

In this context, a first step is to analyze the situation of Argentina with respect to the climatic conditions that can favor the development of forest fires. In this sense, the Third National Communication of the Argentine Republic to the United Nations Framework Convention on Climate Change (3CN)[36] gives some elements to analyze this situation. Firstly, the increase in temperature: In most of continental Argentina (with the exception of the Patagonian region), the increase in temperature between 1960 and 2010 was up to half a degree centigrade, with the salient feature of greater increases in minimum temperatures than in maximum temperatures (which had generalized decreases in the center of the country). In Patagonia, on the other hand, the increase in temperatures in the same period was greater, even exceeding one degree Celsius. In this region, unlike the rest of the country, the increase in the maximum was equal to or greater than the minimum.[37] Secondly, the evolution of rainfall: Although precipitation throughout Argentina shows strong year-on-year variability, there are clear trends for the period analyzed. In fact, for 1960–2010, precipitation increased in almost all of the country, although with interannual and interdecadal variations and differences between the various regions. The eastern zone (the Pampean and Chaco plains) showed the greatest increase (up to 200 mm in some regions), but the highest percentage increase was in the arid zones. This circumstance implied a greater possibility of agricultural use of the soil, which in turn (in the presence of widespread technological packages and international price incentives for the products that can be exploited), resulted in a strong shift of the agricultural frontier that, in most cases, determined an advance on more fragile ecosystems. However, this behavior of the climate variables was not homogeneous throughout the country since, for example, in the Patagonian Andes and Cuyo, rainfall decreased significantly between 1960 and 2010. Another factor to note is that, in both the west and the north of the country, dry periods have become increasingly longer. In these regions, where there is little or no rainfall in winter, the increase in the maximum number of dry days is indicating a change toward a prolongation of the

[35] The author is reluctant to call "neo" a process that has been repeating in Argentina since the 1930 putsch and "liberal" a way of seeing life that only wants freedom of trade and that has often been complicit in the removal of individual freedoms and has systematically opposed any conquest of rights by less favored social groups. That is the reason why in this context we call "neo-free-trade theory" that what is generally referred to as "neo-liberalism".

[36] See Secretaría de Ambiente y Desarrollo Sustentable de la Nación (2015).

[37] See SAyDS (op. cit.).

winter dry period, which could generate problems in the availability of water for some populations, more favorable conditions for uncontrolled forest and pasture fires, and stress on livestock activity.[38]

In this context, a first step is to analyze the situation of our country with respect to the climatic conditions that can favor the development of forest fires. In this sense, the Third National Communication of the Republic, notwithstanding the presence of these regional and/or local climatic conditioners, some situations of water shortage are related to economic activities that generate drastic changes in the natural environment. In the case, for example, of the Salta's Province Chaco Region (with special emphasis on the departments of Rivadavia, San Martín, and Santa Victoria Este), the clearings originated mainly by the exploitation of diverse "agribusinesses" (intensive agriculture and cattle raising) have been developing increasingly for many years (at least in the last four decades).[39] These activities are carried out by both the old and the new large landowners established in the area,[40] who appropriate most of the available water (from irrigation systems or cistern trucks from nearby municipalities) and are limiting access to water for entire peasant populations, both "*criollos*"[41] and "*originarios*",[42] but mainly the latter. This situation is such that most of the original communities have partial or no access to water.[43]

In this context, a first step is to analyze the situation of our country with respect to the climatic conditions that can favor the development of forest fires. In this sense, the Third National Communication of the Republic, at this point, it is clear that beyond the mentioned regional and/or local climatic conditioners (lack of precipitation in some regions at some time of the year, high temperatures, loss of humidity in the soil, strong winds), there are other causes that can contribute to the outbreak of fires. In the case of Argentina, 95% of forest fires are caused by human activity (fires and poorly extinguished cigarette butts, abandonment of land, preparation of grazing areas with fire, neglect, etc.) and only the remaining 5% are due to lightning strikes and other natural causes.[44] In addition, there is a marked seasonality in the risks of forest fires, depending on the different regions and provinces. Thus, between the months of December to March the Patagonian provinces have a high risk of forest

---

[38] See SAyDS (op. cit.).

[39] Some estimates suggest that in the last 15 years, more than 1 million ha have been deforested in the "Chaco Salteño". See "No son seis meses de sequía, son cuarenta años de desmontes" ("It's not six months of drought, it's forty years of deforestation"), Cuarto Poder Salta (09/11/2019) https://cuartopodersalta.com.ar/no-son-seis-meses-de-sequia-son-40-anos-de-desmonte/.

[40] In that area is located, among others, the "El Yuto" establishment that belongs to the Macri (former president of Argentina) family. See Cuarto Poder Salta (op. cit.).

[41] European descendants born and naturalized in the American continent are called "criollos" in all Latin America.

[42] Original indigenous people.

[43] At least 300 families, from the Departments of Rivadavia, San Martín and Santa Victoria Este, say they are suffering the consequences of the drought. See Cuarto Poder Salta (op. cit.).

[44] See Sistema Nacional para la Gestión Integral del Riesgo (SINAGIR). https://www.argentina.gob.ar/sinagir/incendio-forestal/causas.

fires.[45,46] In turn, the provinces of Entre Ríos, Corrientes, Misiones, Chaco, and Buenos Aires have a higher risk of fires in the period from October to March. While, for the provinces of Cordoba, Catamarca, La Rioja, Mendoza, San Luis, Santa Fe, Santiago del Estero, Tucuman, and all the north of the country, the risk begins with the first frosts of May and extends until November.[47] To make the situation even more complex, there are studies[48] that seek to demonstrate that there may be a causal relationship between economic incentives and intentional forest fires. The increase in the number of these types of fires would be caused by the intention to circumvent the prohibition on clearing native forests in restricted areas and thus be able to exploit those areas, outside of compliance with Law 26,331, seeking the greatest short-term profitability, taking advantage of the sequential implementation at the national and provincial levels of that law. In this sense, the cited study identified an increase in the number of fires during the period 2009–2011, in which the number of fires doubled, on average, from 4 to 8 fires per 100 thousand ha of forest land, compared to the entire period 2002–2014.[49]

Thus, whatever the cause (natural or human), it cannot be denied that the possibility that forest fires similar to those that occurred in other parts of the world could occur somewhere in Argentina exists. For the various reasons cited above, this is a latent risk, but it need not lead to inaction, far from it.

In the face of environmental events of the magnitude of "sixth generation" forest fires, the alternative presented to confront them (and try to minimize their impact) is to hide them with a set of public policies aimed at this end and with a plan for the integrated management of the risk of forest fires. These public policies must address these events in a comprehensive manner, promoting the conservation, restoration, and enhancement of forests, biodiversity, and landscapes. The management plan should focus on knowledge of the causes of fire, on the definition of measures to reduce it,

---

[45] At this moment (February 2020), fires are being developed in the El Doradillo urbanization, near the city of Puerto Madryn. The causes of the origin of the fire are unknown, but (whether intentional or due to negligence) are clearly human. The fire extends to about 20 ha of the mentioned urbanization. The combination of lack of rain and strong winds facilitated its spread. See Ambito.com (03/02/2020) https://www.ambito.com/informacion-general/chubut/puerto-madryn-importantes-incendios-viviendas-danadas-y-evacuacion-total-n5080573 and also Río Negro (03/02/2020) https://www.rionegro.com.ar/incendios-en-puerto-madryn-pidieron-la-evacuacion-total-de-la-zona-afectada-1244028/.

[46] This is not the first time there's been a fire of this kind in the vicinity of Puerto Madryn. It is very well remembered the fire of pastures (with an extension of 6 km of front by 40 km of bottom) that began on January 21, 1994, at a distance of 15 km of the city, that cost the life to 25 voluntary firemen of between 11 and 23 years. Also, at the end of January 2001, about 50,000 ha were burned in a fire that reached 800 m from the southern neighborhoods of the city and started with a cigarette butt thrown from a vehicle. See La Nación (15/01/2019) https://www.lanacion.com.ar/sociedad/tragedia-bomberitos-madryn-historia-25-chicos-dejaron-nid2211147 and also Río Negro (01/02/2001) https://www.rionegro.com.ar/puerto-madryn-se-encuentra-sitiada-por-las-llamas-FFHRN01020120011007/.

[47] See SINAGIR (op. cit.).

[48] See Egolf (2017). https://inta.gob.ar/sites/default/files/inta_cicpes_instdeeconomia_egolf_p_tesina_mae_ucema.pdf.

[49] See Egolf (2017, op. cit.).

on the design of instruments to minimize the negative effects, on guaranteeing the protection of human life and property, as well as the goods and services provided by ecosystems, and on the design of an efficient and effective detection and extinction system. There are examples from other countries that are working on the definition, development, and implementation of these instruments.[50] In a federal country, both the plan and the corresponding public policies must necessarily be articulated among themselves, at all levels of the State (national, provincial, and municipal), and must address both damage prevention and the increase in society's response capacity, so that they can have a scope that goes far beyond merely attending to the emergency once the event has occurred. It is necessary, on the one hand, to "do what is possible to prevent the preventable" (forest fires due to human causes, for example); on the other hand, it is essential to build alternatives to generate conditions of greater resilience for all systems, both natural and human, in the face of the possibility of environmental conditions that facilitate the origin and spread of these fires. Adequate land use planning that takes into account the best possible use of the land and its sustainable management, policies for the conservation of natural forests, productive diversification (which limits the role of both monoculture and the irrational exploitation of resources), can be some tools that help to improve this resilience. Forest fire prevention and monitoring systems play a fundamental role in this articulation. A well-oiled coordination between the different parts of the State, the economic agents involved and the social organizations that represent the interests of the inhabitants is another fundamental pillar of this construction. You have to "get your act together" and work on it. Climate change does not help, but inaction, lack of understanding and the inattention to thinking that everything is "magically" solved by the market does not help.

References

Association of Forestry Professionals of Spain & Spanish Society of Ornithology (2012) Proposal for a new forest fire policy in the Cantabrian Coast. Retrieved from https://www.seo.org/wp-content/uploads/2015/12/Una-nueva-estrategia-de-lucha-contra-el-fuego.pdf

Australia: everything you need to know about forest fires. C/NET, 13 Jan 2020. Retrieved from https://www.cnet.com/es/noticias/australia-incendios-causas-consecuencias-koalas/

Egolf P (2017) Econometric study on forest fires and economic incentives based on the forest law in Argentina. CEMA University. Retrieved from https://inta.gob.ar/documentos/estudio-econometrico-sobre-incendios-forestales-e-incentivos-economicos-a-partir-de-la-ley-de-bosques-en-argentina

Fires in Australia: why they are so ferocious and other keys to understanding what is happening in the country. BBC World News, 6 Jan 2020. Retrieved from https://www.bbc.com/mundo/noticias-internacional-50992270

[50] In some countries, a number of plans and policies have been developed in this regard. A particular case is that of Spain. See, for example: Asociación Profesionales Forestales de España (PROFOR) and Sociedad Española de Ornitología (SEO) (2012).

Fires in Australia affect more than 10 million hectares. Telesur, 13 Jan 2020. Retrieved from https://www.telesurtv.net/news/australia-incendios-forestales-millones-hectareas-quemadas-20200113-0020.html

Forest fires are advancing in Australia (2020) Página 12, 5 Jan 2020. Retrieved from https://www.pagina12.com.ar/240061-los-incendios-forestales-avanzan-en-australia

Intergovernmental Panel on Climate Change (IPCC) (2007) IPCC fourth assessment report (4AR). In: Climate change 2007. Synthesis report. Retrieved from https://www.ipcc.ch/report/ar4/syr/

Intergovernmental Panel on Climate Change (IPCC) (2014) IPCC fifth assessment report (5AR). In: Climate change 2014. Synthesis report. Retrieved from https://www.ipcc.ch/report/ar5/syr/

It's not six months of drought, it's forty years of clearing. Cuarto Poder Salta, 9 Nov 2019. Retrieved from https://cuartopodersalta.com.ar/no-son-seis-meses-de-sequia-son-40-anos-de-desmonte/

Koalas: towards definitive extinction. Página 12, 6 Jan 2020. Retrieved from https://www.pagina12.com.ar/240178-koalas-hacia-la-extincion-definitiva

National Secretariat on Environment and Sustainable Development of Argentina (2015) Third national communication of Argentina Republic to the United Nations convention framework on climate change. Retrieved from http://3cn.cima.fcen.uba.ar/docs/3Com-Resumen-Ejecutivo-de-la-Tercera-Comunicacion-Nacional.pdf

National System for Comprehensive Risk Management of Argentina (2019) Causes of forest fires. Retrieved from https://www.argentina.gob.ar/sinagir/incendio-forestal/causas

Portugal's 'perfect fire': eucalyptus trees, desertification and "fire bombs". Libertad Digital, 20 June 2017. Retrieved from https://www.libertaddigital.com/internacional/europa/2017-06-20/el-incendio-perfecto-de-portugal-eucaliptos-desertificacion-y-bombas-de-fuego-1276601388/

Puerto Madryn: major fires, damaged homes and total evacuation. Ámbito Financiero, 3 Feb 2020. Retrieved from https://www.ambito.com/informacion-general/chubut/puerto-madryn-importantes-incendios-viviendas-danadas-y-evacuacion-total-n5080573

Puerto Madryn is besieged by flames. Río Negro, 1 Feb 2001. Retrieved from https://www.rionegro.com.ar/puerto-madryn-se-encuentra-sitiada-por-las-llamas-FFHRN01020120011007/

Río Negro (2020) Fires in Puerto Madryn: they called for the "total evacuation" of the affected area. Retrieved from https://www.rionegro.com.ar/incendios-en-puerto-madryn-pidieron-la-evacuacion-total-de-la-zona-afectada-1244028/

The causes of the fires that 'suffocate' Australia. Milenio, 20 Jan 2020. Retrieved from https://www.milenio.com/internacional/europa/la-sequia-el-aliado-de-los-incendios-en-australia

The fires are "mutating" and Argentina has "vulnerable" areas. Telam, 19 Jan 2020. Retrieved from http://www.telam.com.ar/notas/202001/425211-los-incendios-estan-mutando-y-argentina-tiene-zonas-vulnerables.html

The Indian El Niño: what is the Indian Ocean Dipole and what does it have to do with floods and forest fires. BBC World News, 9 Dec 2019. Retrieved from https://www.bbc.com/mundo/noticias-50705069

Tragedy of the "Madryn's firefighters": the story of the 25 boys who left their lives fighting the fire. La Nación, 15 Jan 2019. Retrieved from https://www.lanacion.com.ar/sociedad/tragedia-bomberitos-madryn-historia-25-chicos-dejaron-nid2211147

War to eucalyptus, the "gunpowder" of the fires in Galicia. El Independiente, 6 July 2018. Retrieved from https://www.elindependiente.com/desarrollo-sostenible/2018/07/06/guerra-al-eucalipto-el-polvorin-de-los-incendios-en-galicia/

Why the fires in Australia are so devastating and do not stop: three keys that explain the dramatic situation. Infobae, 7 Jan 2020. Retrieved from https://www.infobae.com/america/mundo/2020/01/07/por-que-los-incendios-en-australia-son-tan-devastadores-y-no-se-detienen-tres-claves-que-explican-la-dramatica-situacion/

The manufacturer's authorised representative in the EU is Springer Nature Customer Service Centre GmbH, Europaplatz 3, 69115 Heidelberg, Germany. If you have any concerns regarding our products, please contact ProductSafety@springernature.com

Printed and bound by CPI Group (UK) Ltd, Croydon, CR0 4YY
01/10/2026
02225883-0012